Earth's "Conditions" and Extraterrestrial Reality

Earth's "Conditions" and Extraterrestrial Reality

Frank Mazzucco

VANTAGE PRESS
New York

FIRST EDITION

Published by Vantage Press, Inc.
516 West 34th Street, New York, New York 10001

Manufactured in the United States of America
ISBN: 0-533-09721-5

Library of Congress Catalog Card No.: 91-91076

0 9 8 7 6 5 4 3 2 1

Earth's "Conditions" and Extraterrestrial Reality

1

The following knowledge is not against any form of religion or political inclination, but mostly is a new understanding about human behaviour and the natural world in which we live.

First of all, we have to abandon the old idea that everything we think or do is completely maneuvered by people, but accept the idea that a natural plan, which dwells in the molecules, cells, etc., is the sponsor of much we do either at a physical level or at an ideological level. In other words, it is not up to us to choose our dreams or the colour of our eyes and life span, but mostly all is according to that created plan that has been alive in us and all living things since the beginning of life on our planet. Many political figures, discoveries, religions, etc., were nothing else but part of a natural plan that was and is part of our human advancement, correction, balances. Nothing else but a form and a way of growth.

In curiosity, we want to ask why we have a world with so much pain, grief, war, etc. The real story is that the Creator of this world could not use a magical force and therefore had to create a mechanism (nature) instead. A natural mechanism is like a motor. And like a motor, it is our adaptability for self-automotion that makes us succeed. Again, in curiosity, we can now all ask why then the Creator of this world did not create a human world where our prehistoric ancestors could understand everything right from the very beginning? Why did it take so long for our discoveries and to have a more humane society? Well, before the accomplishment of intelligence, a body was needed first.

Why did it take so long for the cures of disease to be discovered? Well again, if magic could not be used, then an advancement could only come across a long period of time and not certainly in the length of one day. In a not magical world, in order for our vegetables and animals to grow and produce it takes months, years, across stages and processes. It has to be understood that because everything is sponsored by nature, maybe we have to accept the idea of taking the world as it is, with its pains, wars, sadness, political bad inclinations, diseases, etc. This is untrue, even if what happened in the past was a natural factor to be in our human history. The fact is that Mother Nature can be restyled, grow forward, if we can make things or ideas according to workable conditions or solutions. Proof of growth is that our ancestors used chariots and today we have automobiles, trains, airplanes, and spaceships. That tells us that Mother Nature offers chances for remodernization and growth not only on a physical level, but on ideological, emotional, and spiritual levels as well. On the physical level, we have, on our planet, made a remarkable advancement, but in our ideology, feelings, and spirit we are mostly unsettled and human behaviour is mostly still at large or astray. Our ideological images have remained almost unchanged across the millennia or since our ancestors, way back in history, left the forests to become cave dwellers. It is very obvious that religions, political inclinations, discoveries, etc., are a natural process and it is understandable that no one can change the world completely; this is not possible, because our world is not magical, but a nature-matter plan. If we have a plan we can follow and help that natural process and all religions, ideologies, political structures, feelings, and spiritual images, we could be advanced to a more complete stage for our contemporary level. This is an existing possibility because our world is created in a cycle and for advancement in a physical and ideological structure there is possibility for "remodernization" above all ideas, emotions, and ideologies of the past human history. This way we help Mother Nature to advance if we obey her according to her logical plan, and this will certainly benefit humanity.

Before any further comment or explanations, it is opportune to understand why diseases, either physical or mental, etc., happen to be on our human scene. Diseases, either mental or physical, come or happen because we accidentally break that physical and ideological natural central geometry. What do I mean by a natural geometry? Well, in our world we need a centre line along which all things can be set to form a mechanism. Without a centre line it could not be possible to form any living thing and any mental direction and therefore everything could be on the loose, or unassembled. How does this geometrical line or centre line come to be in plants and us? Vegetals obtain a geometrical line by molecular involvement with the pattern of the sun in the sky, which forms a half-circle from sunrise to sunset (in reality the earth rotates around the sun). This half-circle is the principle for all geometry in living things. The phases of the moon and its pattern extend the basic geometric line. That geometry is essential to give us a centre line or balance and laws for our body and ideological structures. This is why we have laws to follow, which are also the principles of our human ideological automotion "within limits." Flowers and leaves have a design with a centre line. We as humans are based on differences, with the same structures in geometry, but no magical factors; therefore anything off-course or off-centre from that central structure will be the cause for a mental or physical problem. It is too early in my writing to explain geometrical things in full, and so it is better if we begin to search away back in the prehistorical era and understand how and where our ancestors began their off-centre behaviour, keeping always in mind that whatever has happened across the many, many millennia to our human race was the only process and alternative to bring us forward as we are today. It is not essential that we follow that old pattern, because there are possibilities for changes, growth, and recycling from the bases.

Most of us today know that our ancestors used to live in the forests, using the same behaviour as any animal then or today—little intelligence, little knowledge, helpless in many situations. But things began to change from that natural process created and our ancestors began to be aware of the surrounding world—the use of force against the weaker man in order to have the best meal. We see this in the animal world today; the strongest always survive and get the better portions of food. Millennia passed and worldly things did not stand idle. Our prehistoric ancestors now had the knowledge to use force in order to obtain the best meal in their small community. This was the first realisation of using force to acquire needs or wants. The second realisation was the discovery of using force as a tool to intimidate the weaker man to supply food for the stronger. These two first developments of ideas of our prehistoric ancestors are the *basis* of our principles all over the world today. The strongest of the tribe or community became the leaders, thus paving the way to form a monarchy in the centuries to follow. This tribal leader (monarch) was a raw ideological image but very essential and attractive to our early ancestors to break away, experience, and explore a way of life different from before. This was certainly different from the rest of the animals in the forest and other living things. From this knowledge came the development of hunting, killing, and the use of wildlife for games, and as centuries passed the wildlife decreased sharply. Animals developed the instinct of fear of man and in turn made it difficult for the hunters. This fact and others forced our ancestors to look to a second source of food—agriculture.

Agriculture was another turning point in our historical scene. The cultivation of plants was much more strenuous and less challenging than hunting. Our ancestors had the knowledge to use force and intimidation to cause the weaker tribesmen to do work in the fields. Now slavery began, a behaviour carried from one generation to another based on old ideas according to their knowledge, since they were cave dwellers. Millennia passed; there were fewer forests, fewer animals, and agriculture was advancing. In the natural evolutionary process, our ancestors had to advance intellectually as well. So if magical factors could not be used, what other choice did Mother Nature have but to bring us forward intellectually? Mother Nature is a mechanism, she works on its own atomic or molecular balance, so the life-style of our prehistoric ancestors who were hunters/cave dwellers and the second life-style of the agricultural civilisation suffered a gap. This gap can be described as a change in molecular structure of the food chain—natural, fresh, hunted food with wild fruit versus the diet based on a long-range planting cycle plagued with weather conditions, inexperience, and pests. This imbalance resulted in human degeneration of that genetic blueprint that had been functioning, always the same, in the forests for millennia and now because of the advent of agriculture created differences, and these differences caused new ailments. Diseases caused pain; pain aroused wonders; wonders produced discoveries; discoveries produced cures.

The knowledge of our ancient civilisation was very limited, so our ancestors attributed their pain and hardships to supernatural causes. This led to fear of the supernatural, and they began worshipping false idols and gods. This did not cure the ailments, so ultimately they looked towards the human body for answers, which did centuries later lead to the discovery of the human body structure—the heart, cells, molecules, atoms, D.N.A., bacteria. All of this natural process of understanding, until today, is mostly at a physical level only. As our world is physical *and* spiritual then diseases cannot be looked at only on a physical level, and so we will be forced to understand how the mind operates. Our natural mechanism cannot instruct us directly, because the natural process is a mute system of molecules and cells. The past brought us imbalances that led to wars and ailments, and mental problems will increase in the future until we understand prevention. Problems we must deal with on the ideological level are: war, pleasure in sadism, revolution, rape, incest, exploitation, abortion, depression, superiority and inferiority complexes, murder, suicide, emptiness of spirit, and the list goes on. People becoming frustrated with the increase of problems will push people to look to the mind and the cause that produced these problems. Thus we learn of prevention, as well as cures. With the mental level, we have preventative methods and instructions. These are the cures at a spiritual level. This means that all world religions and ideologies in the years to come will have to play a very specific, important role in order to fit that proper mental mechanism. Those important and calibrated religious ideas' important roles will be a main prevention of and antidote to small and large mental abnormalities, but religions around the world must be remodernized to be a workable tool. In other words, we have to understand how our minds work in order that in the future we can use proper and complete knowledge to shape the best course to follow for humanity in the centuries to follow. When our knowledge of everything is completed we can use ideologies, recycled, so things are or will appear always fresh from one generation to the other.

Life, then, cannot be complete with merely a physical understanding. We need to know how a mind operates; feelings must be understood and so, too, the spirit. We didn't understand Mother Nature properly, and that produced disease in the human race, which led to the understanding of the "easy parts," that is, the physical level. Once we understand these problems, it means that we have to apply good ideological methods, meaning that all world religions and political systems in the years to come will play very specific, important roles. Those important and well-known scientific, religious, and political roles will be a main prevention antidote to small and large mental abnormalities, if applied for good reasons. But religions and ideological systems around the world must abandon their old and rudimental pattern and be remodernized to be a more workable tool. In other words, as I said, we have to understand how our minds and feelings work in order that in the future we can use them (ideologies and religion) directly and recycling in complete knowledge for that sane course to follow for humanity in the years to come.

As I said, at a physical level we are well advanced. We have almost full knowledge of all our organs in our body to the cells, molecules, genes, and so on. We have developed machinery, computers, televisions, radios, telephones, etc. We have the knowledge of the solar system. But on the ideological and spiritual side we are almost at a "raw beginning." This explains clearly that despite our technologies, comforts, and economy, we have just too many problems at the ideological, mental, emotional, and spiritual levels.

As stated before, we have a created system that lacks magical power. It only enabled our nature to begin growing not according to a desired process, but to the existing conditions offered in the immense space that we see and the elements or matter on earth that we know, this being reason enough for our ancestors to move from forests to caves, then, later on, create civilisation as we now know it. Mother Nature had to follow a determined course related to conditions, and that process resulted in a well-developed body structure, adaptable to climbing, searching for food, building, etc., but our ancestors functioned in a very basic and primitive manner intellectually. Due to this fact, it took many thousands of years to discover a cure for many of our present diseases. This tells us that our human natural advancement develops in stages, somewhat like our school system of elementary grades, high school, then university. A high natural process or university level will enable us to understand how our mind came into being, as well as mental and emotional dilemmas and, finally, "Spiritual" images. When humanity understands and accepts this complete knowledge, then they can choose a safe course that will be beneficial for all of mankind. Until then, we work with limited knowledge, causing history to repeat itself in the painful old way. If we fail to understand and explain that integrated natural process we will not have safe advancement.

Our body lines and contours are based on a geometric system, obtained between food chains, adaptations, necessity, etc. The chemical elements are the materials to make the human body possible. The resistance of substances in our body can only support determined conditions in terms of physical and mental-emotional weight. To depart from or abuse that natural standard plan can result in an overweight or "lump" from that physical and mental standard centre, going "off-course," leading to the causes of the many world problems of today.

The mind, our feelings, and emotions are not classified as solids or substances or physical, but are more appropriately considered as spiritual forms. Medicines belong to the world of solids or substances and cannot be considered for a cure of the spiritual and mental levels. Medicines can make the human cells strong or keep them in balance and help them to fight intruders, etc., but healthy cells can register an idea, an emotion, and feelings not aligned with natural possibilities causing sadness. Nowadays because of the impossibility of understanding our world we have a mental and emotional (also physical) "environment" too poisonous and dangerous for our own existence. Our human body is not magic, but within limits on its physical and ideological levels. If we could master all natural laws and their possibilities, we could also find the keys to peace, basic prosperity, good feelings and health, real friends, and political ideas, the principal ingredients of a good spirit. If we then can succeed in having a normal physical life and ideological life and a good spiritual life, it is then obvious that there are cures at any level. For physical ailments one can get prescriptions, but for the mind, feelings, and the spirit the healings are mostly preventative and ideological. Mental and emotional problems can only be prevented if all world religions and ideologies become remodernized in their beliefs, according to that true natural plan. Bad feelings and mental unrest can develop in the same manner as one catches a common cold. Common causes can be the contact with spoiled people with personal thoughts gone astray; misunderstanding, misguidance and nonacceptance of the certain nature of things; acts of not being in conformity with our own personal environment; or ideas can create the ideological virus (contrasts). Mother Nature offers conditions to limit offences and abuses if we adhere more or less to that centre line of the natural process, and from there we can limit the radius of expectation, acts or applications at the physical and ideological level.

We must not fall into believing that our present system can never be changed, or we not only mistrust the natural process, but we also misunderstand it. If we do not understand it yet, it is only because our Mother Nature, as the created plan, couldn't begin our human civilisation everywhere at once but gradually or in steps or in a form of "growth." Our world is hard, difficult, and tricky. In the physical sector in our human body, we can use medicines, which when applied work in most cases very well. The present problem today is at a mental level. Our mind development does not use the same process as the physical combination and thus cannot be cured by physical medicine alone, but we must not despair. There are preventative "ideological medicines" that can work well if followed accordingly. In the physical world we have proteins, vitamins, calcium, etc., which in turn form and sustain the human body. Now we have to depart or differentiate from the physical world and think of which way, in an ideological form, we can use the proper vitamins, proteins, (various ideas) etc., keeping always in mind that our Mother Nature uses the same laws, but differently. Today, as well as in the past, cures at any level are always present, but not discovered. For example, penicillin as a cure for infection could have been known thousands of years ago if someone came up with the idea, but it was discovered only recently.

So, too, this applies to the mental and spiritual ailments. The physical level is a visible process, but the ideological level is an invisible process. Thus discoveries of cures or antidotes are difficult; therefore, many good ideas, religions, political ideologies, etc., have been put aside because of the difficulty of understanding. Again, we say vitamins and proteins are to the physical level as religion and political proper inclinations are to the mental level. The abuse or lack of them can lead to pains in the feeling part of the human body system and to small and big mental problems.

D.N.A. is an engineered process for the physical body structure. When religions and ideologies will be fully understood, it can become a perfect D.N.A. for the mind, feelings not only at a personal level, but in genes of the fetus during gestation and into the future generations. Prevention of mental and emotional problems can indeed be a realisation. Now we can at last feel that the cures are possible for the mind and the emotions. We have to discover what words, actions, and "image" works best for our mental perfect D.N.A. and to build a logistic future for our heirs. For a first step we have to go back to the prehistorical era and up to today and understand what caused our ancestors' precarious knowledge to deviate from that central ideological natural course. Imagine now our prehistoric ancestors. Physical force, hardship, pains, and discomfort created the development of the ideas to allow us to be in less pain and more comfort. These ideas grew and became the tool to use force at will. This force was the wrong method to obtain the best meal for the group, for example, because the idea of sharing should have been the tool used. This was the first deviation or going astray because missing from their knowledge was the word *sharing*. The concept of sharing was absent; the ideological or mental process suffered from one era to another, and this had painful consequences. Examples of pain are slavery, wars, assassinations, conquests, tyranny, treachery, etc. "Sharing" in the animal world has developed very little and is still very simple and basic. The stronger of the group gets the most food. Humans, on the other hand, share among their own kin, but many times they are unwilling to share with other people, races, or nations. This shows that we have not progressed in full from the dawn of civilisation. It is time now to advance in our ideological process.

2

The main factors in the development of ideas at the beginning of man's civilisation were force, hardship, and life difficulties. Force, ideas, pains, and fair and unfair treatment developed the feelings and emotions. Force, emotions, curiosity, feelings, and pains produced intelligence. Intelligence opened the doors to "reasoning" and from here led to many discoveries and to the summary civilisation that we have today. We share the world with many other living things, and as they are involved more or less with that same natural law that we humans share, then if we study them we will have a better understanding of ourselves as well. For example, let us observe a flower. It is composed of many parts, telling us that the molecules involved in the unfolding process of the flower are diversified because of the involvement of laws and various elements. Petals, pistil, stigma, ovary, etc., are examples. But the "whole structure" of any flower does something unique—it opens to the world to only one form of light: *the sun.* Compare this development of nature to our human ideological unfolding, which is diversified by various ideas, but globally we believe in *only one God* as that form of spiritual belonging.

Our created world is a mechanism, made of many parts, and keeping people under the same ideology can be wrong. People vary from generation to generation, because of their learning, actions, hardships, and enjoyments, so the genetic background of each person is different from every other person (except for identical twins), and logically, people cannot be attached to the same ideology.

Differences in people stem from attachment to things, laws, emotions, goals, ambitions, and religions. Therefore, each individual must choose an ideology concerning goals, emotions, education, and religion that fits the mind and body under present conditions. If we could understand this process, our human world could have existed in peace millennia ago.

Our human ideological world is a natural process within limits, but this ("within limits") has been much misunderstood, generating confusion and bloodshed. Ideas are the leading stars of the physical body, which enable an individual to be comfortable to do his thing. When that natural proper ideological comfort is contrasted or twisted, the individual will feel insecure and unhappy in doing the usual proper routine. Certain individuals, for example, can regroup, sharing the same laws, with moderate ambitions, concerning religion, education for the young, etc. While other individuals with high or different ambitions—such as can be seen in attachment to wars, hard work, luxuries—should form another group separated from the first, but logically belonging to the same nation, just different in ideas. For a further understanding and comparison to us humans we must study the plants and animals. We share with them the same world. After all, without them we would not be here. In the animal world they live constantly among challenges and danger. Failure to be alert means they fall prey to the enemy. Animals kill for food. On the other hand, the plant world has a relentless challenge among species not only in getting nutrients, but in getting the best spot for sunlight. Grass has difficulty growing where trees are predominant, so it is true a tree seed has difficulty growing where grass is predominant, because the healthy, thick leaves and roots of the grass absorb light and nutrients from the soil, this takes away from a seed or young growing tree. Without the process of photosynthesis, this same seed becomes weak and prey to insects and disease, and these are the battles or challenges among plants. Many plants produce poison to get rid of other plants and animals. Some produce fruit in abundance, as they know they must feed a multitude of mammals and insects, so some uneaten fruit will reseed and multiply. Many plants have spines to defend themselves against herbivorous animals. Ivy, for instance, winds itself around trees, making it quite impossible for leaves of the tree to take full advantage of sunlight. Lack of sunlight makes the tree yellowish, open to disease and pests, and it sometimes dies as in a slow strangulation in terms of light. For a comparison between animals and vegetals we can call the ivy plant the "boa constrictor" of the forests. Even the fragile mushrooms have a way to succeed and live, because they grow in the dark, not needing the sunlight to produce chlorophyll. It seems that mushrooms know that animals don't roam at night, thus they will not step on the fragile growing structure. All this should make us wonder and think that there is much know-how in the plant world necessary for their survival.

With this we can draw the conclusion that intelligence does not exist only for mankind, but is part of that natural process available for all. The point that I want to make is if plants could each have their own share of light, there would not be the necessity for competition. When cultivated, fruitful plants have enough sunlight and are well spaced from each other, they are healthy and productive. So too if people group with others who share the same ideology, religion, ambition, education, etc.; could we not also be very peaceful, productive and healthy? The plant and animal worlds lack elaborated intelligence and that is the limit to which nature can succeed in them, but we as humans are more aware of the world we have, can avoid all the friction and become more communicable and peaceful.

Differences are the basis of life's success. Agriculture was the main source of food after the eras of hunting in prehistoric times. The utilization of agriculture was, largely, the main attraction of our ancestors, but to advance into more and more civilisation things had to be different from that main attraction of everyday agriculture. More differences are needed, such as the ugly, the beautiful, the ambitious, the less ambitious, tall people, short people, doctors, patients, and industrial and agricultural workers. These differences were the basis to advance past the agricultural level. It could not have been much of an advancement for human life if everyone only worked in the agricultural process and were farmers, for example. We are not in a magically created world. It is understandable that in order to need a doctor, there had to be people with ailments. Being a doctor provided a form of life-style and a different way of life from agriculture. Another life-style other than agriculture is that of a police officer or soldier, which is different from being involved in hunting and agriculture, but unfortunately there has to be delinquency and wars. To become rich is another ambitious diversifier for goals, but to become rich and successful there must be workers, the poor, and less intelligent people.

Our creation since it all started back in time in terms of stars, planets, and life in general as we see it on our planet has always been a very methodical and calculated, intelligent process. Certainly we cannot search the stars for answers because of the distance, but rather we have an easy and accessible world of animals and plants to study. We must first of all focus on animals' and plants' "survival" and accept their competitive terms in relation to their limits for study and understanding, but not fully applicable to mankind. We can use our intelligence in a more extended radius than the rest of living things. In the vegetable world there is always a constant battle among plants in order to obtain sunlight and nourished soil for themselves and for the seeds to grow. If plants are contrasted in that gaining of proper amount of light to develop their molecular structures, they (plants) grow tall and thin, abandoning that process of certain molecules that are employed to form flowers, for example, and can be deviated to grow other structures of the plant instead. This, for the vegetal world, is a normal alternative for their survival, as they lack the intelligence to verbally request other trees to move over and stop casting shadows.

In our human ways, for a comparison from above, very often we are pushed away from the central line of normal growth because many individuals are too ambitious for themselves, leaving not much chance for others. When an individual born "normal" is overruled, in contrast, he begins to be psychologically depressed and his duties to good causes begin to be lost, as in the case of a plant when, in a competitive race to get sunlight, it will grow tall and thin and has to abandon the process of flowering and forming fruit because there is not much left for this. On the other hand, when an individual's intelligence misunderstands our created world, he or she is then misled into ideas, situations, and ambitions that are difficult to fulfil, and here too things can grow into depression and apathy towards the world we have. In the extreme end, many individuals use tricks, physical and psychological intimidation, to create an attitude of success in high ambitions, not at hand for the vast majority of people. This creates a sense of failure for the less successful people.

This also leads to depression, revolutions, ailments, commotions, and wars. From the natural point of view all the way from the beginning of civilisation, everything is, in a way, normal, as nature gives us a chance for advancement or growth then not everything can remain forever according to our present and past times. Taking for instruction the vegetal world as the nearest point of information, we can easily see that when a plant grows under too much shadow from other trees, it grows yellowish, and produces fewer flowers and fruits, often leaves fall prematurely, etc. In our human world when there is too much difference from one individual to another in terms of wealth, power, job position, taking and receiving commands, etc., many will get depressed, silent. (In vegetal terms, yellowish means that there is something wrong in the structure of the plant that is much like depression in humans.)

In the vegetal world a small, fruitful tree becomes fruitless when overshadowed by other trees. In the human world, a hard-working person suffering due to "unfair" behaviours and ambitions from others can become "mutant" and can gradually turn to extremes and become a delinquent, cursing God and the world, finding it difficult to speak to friends, bringing hate and misguidedness to the family, uneasy with religions, putting God's rightness into question, and so on. Plants use a form of poison to eliminate other plants. Humans use physical and ideological evils to subdue others to their ways. But plants and animals seem to not create much contrast when in contact with and in the presence of members of the same species. From this natural condition we can learn something for our human benefit. Suppose our human race grouped with other individuals who shared the same ideology (there should be seven main world ideologies), work conditions, religions (seven main world religions), and ideas on education for future generations, could we then abandon the ways of hardships, wars, delinquency, competition, and ailments? If plants could understand and leave enough space among themselves in order to get proper sunlight, there would be no need for them to fight among themselves. We cultivate many varieties of plants, and with our giving them proper soil and sunlight and distance between each other, they seem to grow very healthy and fruitful, but transplant these same healthy trees to the forest—and, of course, competition—and they become unhealthy and fruitless. According to the above natural information, we can regroup into any one of these seven ideologies acceptable to particular mental strength, body strength, and will. Other people can form their own groups in agreement for high ambitions, if this is their will, and can choose certain religions or no religions and ways to provide things for the community. Groups of people by living ideologically separate will not overshadow or undermine the others and the need to fight in competition will cease to exist and many ailments can be avoided as well. If a competitive ideology is chosen, competition will remain as that reaching goal to satisfy those having success, but also be aware of the danger that may arise as the unsuccessful will tend towards delinquency, easy money through pushing drugs, and can develop apathy and stress as well as experiencing frustrations, suicides, etc.

3

We must always be aware that our human body is a machine and the only way it can adapt to our world is by our giving it a workable, intelligent environment at psychological, physical, ideological, and religious levels. It is hazardous to think that we could avoid or "not let things bother us" when we have an ideological, physical, and religious environment that is not aligned with specific needs for our organism in its complete structure as mind, body, consciousness, emotions, etc. A motor of an automobile is built to run properly if kept accordingly, but will cease to function if we break the valves (rules). For example, our human body structures from the natural assembly should run well but will also cease in part or as a whole to function when purposely or accidentally or by misunderstandings we do not give proper maintenance on physical and ideological levels.

We see that all living creatures in this world have learned things all by themselves, forgetting that who really does the job or has the "know-how" of things (plants, mammals, insects) came from the molecular apparatus that all together makes the whole structure (the body) move and behave. Intelligence of things is then a molecular physical awareness. For example, an ivy plant develops an abundance of leaves, which makes it impossible for the tree to which the ivy is attached to exist, without sunlight, and in many cases that tree dies out. Is this ivy a "killer"? Eventually not. Because the purpose of the molecular apparatus of the ivy is to rise to the treetop to gain sunlight for its own survival, it seems to look like a "killer." The only way to get this sunlight is to use other objects to rise towards it, whether they being trees, fences, or walls. So the necessity of light makes the ivy in this case use intelligence to kill, as in reality it is a molecular necessity to use as much light as possible.

Across millions of years, all the vegetal molecules have accommodated to the existing environment, weather conditions, changes in temperature, defences against the enemies, etc. Any changes to that environmental condition were set-up for a more complex molecular awareness and from here the building of defences, resistance to dry and cold climate, and an ingenious way to disseminate seeds away from the mother plant for reasons of better soil and light. Since the beginning of life on our planet our human body has been an interacting accommodation of molecules to succeed among ourselves, with animals and weather conditions. Any changes and appearance of new elements provoked changes in the body's molecular structures and from here to the birth of new organs, immune defences, adaptability, etc., across millions of years of prehistory. Our body organs then didn't magically appear on us, but because of the specifics of different molecule combinations, which also built different organs, etc.

For information regarding our understanding we have to go back in time briefly to the very beginning of life on our planet. Billions of years ago, after the cooling process ended, the elements were ready to interact with themselves, as well as with sunlight, plus moisture, generating the first photosynthesis and the first molecules towards a future vegetal world. As vegetal molecules were present in the earth's crust, other elements or molecules began to interact (or eat), with those existing vegetal molecules generating plant-eating cells. That primordial soil then had vegetal and animal molecules. Another combination now was possible by the molecules of the elements interacting or eating animal molecules and from there forming the primordial carnivorous cells, much later to become carnivorous animals. Weather conditions across a long period of time, locations in the planet, heat, cold, and so on diversified many of those primordial unicellular creatures, generating the origin of the species. In our human body system the appearance of the organs has been more or less a similar process. The appearance of the stomach, for example, paved the way for the intestines to originate. Stomach, intestines, pancreas, liver, etc., seem to belong to the same family, as a lemon, orange, and grapefruit are in the vegetal world, speaking of the physical process.

Back again now to the origin of human civilisation and the advancement of prehistoric man. The use of force paved the way to intelligence (speaking now in an ideological process). As intelligence progressed, a "motivation" had to appear in order to inspire man to abide with a raw form of laws and provide food and goods for the community. That first motivation and ideology was monarchy. In our world one thing leads to the other. Monarchy alone was an ideological interest to be attached for provisions and laws, but not enough to give birth to the feeling of love, elimination of slavery, consideration, mercy, understanding of others' human rights, dignity, an afterlife, etc. Therefore the appearance of religions was necessary on the human scene. The reasons our civilisation advanced from six thousand years ago up to today were mechanisation, ideologies, and religions. Today in terms of mechanisation it seems as if we have reached a good point, but ideologically and religiously it seems that we are far behind. Our prehistoric man had no feeling of love, consideration, mercy, etc., because only good religions can stabilize that in our body to become good emotions and beliefs, in turn maturing into a good human spirit. An ideology only selects laws to obey and to have food on the table but isn't able to keep our good sense in our feelings. The underdevelopment or lack of proper religions means an "assassination of good senses." As we mentioned, our civilisation rose because of ideologies, ideas, mechanisation, and religions. Take away one of those pillars, and we will fall back in time. Take away religions and we begin to be senseless towards the ones who may be sick and dying, will not care for the poor or weak, or exploit any situation for profit. Divorces will rise and cohabitation will replace marriages in the church. Young generations will come or be born with strange behaviours and will begin to like the emotions of confusion, wars or war movies, and horror movies. Senseless murders will rise. There will not be much coexistence and understanding between parents and the rest of the family. To gain material ambitions, people will use other people as tools to reach their goals. Because of the lack of love and brotherhood in the community, people will not fear to pursue the ways of physical and psychological exploitation as long as there are chances to gain easy money and good working places.

Let's take a tree for an example of religious and ideological cutbacks. If we cut the tips of the branches that have grown this year, then we return that tree one year back in development. If we cut the branches of that tree, we return that tree four or five years back. If we then cut short all the branches and the trunk, we return that tree to the beginning of its existence. If we remove religion from our human scene, then as centuries pass the good feelings will slide us back to prehistoric man, where any sufferance passed unobserved, not understood or explained. On the ideological side, it seems as if we are returning back because we are eliminating varieties of ideologies in the attempt to obtain only one ideology that is suitable for all. This is not possible; not all people have the same mental and physical instructions to follow in unison. An ideology can be excellent and real, but people have been astray for too long, making it unsuitable now for proper ways. This natural world can accommodate to us. There are seven principal forms of ideologies, and each individual can choose one of them to carry on in his possible ways acceptable to the mind and physical strength, as well as agreeing to form laws convenient to his reach. To build a number of new ideologies we need a lot of information from Mother Nature and then we can remodernize the old ideological thought as well as add new and advanced ideas. Before we remodernize, we must understand nature, with its molecular laws involved.

With the birth of ideas in our history of civilisation many psychological diseases also began to enter our human scene. Psychological diseases paved the ways for many of our body organs' ailments. Emotions such as feelings of hate, depression, high ambitions, inferiority and superiority complexes, sensitivity, etc., can overrule that standard pace of nature when contrasted with others or by self-producing an overworking of human organs, especially the brain and from here ideological malaise. An unfit ideological condition creates worries. For worries to be present in our emotional structure they need constant thought to keep them alive. Any physical and ideological constancy that strays from that natural central line can put into confusion or molecularly interfere with any organ or function in our human body. Stranded molecules can interact to form cells that are away from any D.N.A. structures belonging to our human body and can indeed damage, put out of condition, or terminate the existence of it. Our body as a mechanism cannot function properly or at all when molecules differ from that blueprint or D.N.A. that for thousands of years has shaped and kept our body as we have now. Our natural world always offers the same conditions, summers and winters, mornings and nights. Animals and fruits are always of the same design and produce the same vitamins, proteins, etc., and then there is no way that our body's D.N.A. will be terminated by nature, but perhaps by human disobedience and misunderstandings or by wars and artificial ways.

The dangers of using too many medicines and unnatural methods to grow plants and domestic animals can change that nature's central line gradually, and we can be affected. On the other hand, we have ideologies and ambitions that are not able to satisfy all, resulting in worries, depressions, and other ailments. No one at this present time seems to understand religions. With this, much of our proper human good feelings are absent, facilitating the way for more emotional or psychological spiritual ailments. Pollution can change weather conditions and thus affect the vegetal molecular structure, raising new conditions for living things that perhaps will be beneficial for no one. Across food chains we must always be aware of the fact that any change can produce a new awareness for molecules and those combinations will not always favour us. Nature, in order to limit pollution, etc., has the nations develop not all at the same time, but at the present time it seems that all nations want comfort and do not care to limit population. As a result we may have much food and many comforts but are too depressed and sick to enjoy much of anything. Comforts are good, but as we have a nonmagical world we must keep proper nature in order to keep our body and all its organs sane, so they can give us the joy of feelings and the joy to live. All living things are in existence because of that created plan in Mother Nature. As natural laws can be worked out to become more suitable for us, we can choose the good part, or we can keep our way of life as it comes, now with wars, ailments and ambitions, etc. We can have a competitive ideology where the more strong, intelligent, and coldhearted will survive better and will live more materially comfortable lives than others. Or we can elaborate natural laws to a more intelligent state for human manner with modern ideologies and spiritual factors that will accommodate a better will to human survival. The animal world has no intelligence for mechanisation, ideologies, and religion, and for them their world ends there as it is. For us things can become more and more interesting if we apply reason, proper feelings, and determination. With applications in conditions towards a natural central line we can succeed with less commotion and pain. Not in all cases can the use of reason satisfy us and make human existence happy, and we can understand that they (reasons) have a limited extent and we must use religion to make reasons more effective. For example, we can explain the reason to die as to make room for the new generation—otherwise there will not be enough space to accommodate everyone that is born on earth. This reasonable explanation doesn't satisfy our spirit. There are reasonable explanations why we get old, as there are reasonable explanations why we have painful problems, but again that will not make us happy either; our spirit is then suffering.

To enforce our reasoning, to smooth pains, etc., Mother Nature has counterbalanced with the birth of religion almost since prehistoric times. Religion is a tool to reinforce reason. Imagine digging a deep hole in the ground with our bare hands. It is difficult, but by using a shovel or some type of mechanisation things become much easier.

Religion is then a useful, powerful tool in the hands of reason. A good religion is the wings of reason. We have the illusion that ideologies alone can make our spirit happy, but ideologies are bundles of laws as a reachable goal for the followers in order to gain things or to curb infractions, as those are material conditions that make them satisfied in the achievement. Ideologies do not have any possibility of curbing our sad spirit. In this world we will one day get old and die. We can be rich, intelligent, and powerful, but in between or in the end perhaps we will feel miserable without any proper beliefs in the good ways for the spirit and in an afterlife. In any advancement there are needed more and more parts as material and ideas. For example, when the first cells were born relating to the vegetal world, billions of years ago, there also rose to the scene the animal unicellular world. When the first ideology (monarchy) was born, so was the first form of religion born. Everything was in natural balance, and nothing was man made. Absence of intelligence in the past millennia has created big confusion at the religious and ideological levels. History is full of contrasts. Even when religion was applied in good faith it often resulted in bloodshed, but the belief has been kept alive across centuries. Back in time, ancient Egyptians were so sure of another world that they left food for the dead for their journey to the underworld and slaves were killed to go to the next life with their owners. Ancient Romans worshipped idols to obtain their victories and overcome their pains and defeats.

Confucianism, Buddhism, Hinduism, Islam, etc., have their own stories as well. The Christians in ancient Rome were such believers that not even persecutions and crucifixion on the cross discouraged them. Thus this tells us that natural plans are on the human scene and for very good reasons, too.

4

Confusion still reigns over ideology and religion. We need to remodernize many of our ideas concerning ideologies and religions for the spirit and soul. A religion that teaches the existence of one only God and the existence of an afterlife will not need remodernizing in this area. A religion that does not teach an afterlife, God, etc., has to be remodernized in proper structure. Otherwise it undervalues the extension of reason and will put everything in confusion, contributing to the damnation of feeling that we live in a meaningless world. We must understand that without believing in God, Creation, and an afterlife, our own existence and universe become unexplained and thus meaningless, with no basis to start. The use of any religion as a tool to receive magical favours from God, to be rich, or to be healed even when we behave wrongly must be restructured into believing that religion is only for the spirit (and soul) and not for material gains.

However, good faith can cure many emotional ailments and from here may lead to cures for the physical body. For material accomplishments we have ideologies as agreeable laws to limit (or not) gains and ambitions. On the other hand, ideologies have to be restructured and be adopted by various believers. It is unreasonable to apply one form of ideology in the world or in the nation, because people's inclinations, wills, and energies are not the same. Therefore, we need ideologies to fit in normally, and sometimes not, for all races and colours in what they wish to believe and follow. Eventually, before these ideologies can appear, there must be scientific understanding and not just unfounded ideas and propositions given to the people. Scientific understanding from Mother Nature means what is really belonging to morality and normality or, if people want to see normality (or not) as working conditions, ambitions with limits (or not) in school education and teaching, instructions for future generations on acceptance of abortion and divorces (or not), cleaning and protection of the environment, controls of pollution, amount of work and ambitions, morality, the type of acceptable religion, working under supervision or being trained from a young age in schools to work without supervision when in the work force. Eventually any young generation once they reach adulthood will have the chance to choose the ideology that fits their dreams and conditions and can have the possibility to move in places in the nation or the world (if agreed upon). If we could have one ideology to fit all people, mankind would be better, but as we are not a magical creation, then "differences" are the only things that makes our existence possible.

Let's search inside our human body. We have a heart and a brain, which has emotive feelings. It can be possible to understand the wrong behaviour of others and ourselves. We can store hate, and this causes us to kill. Our stomach has an affiliation to the brain and reacts when we see something ideologically and physically vulgar, which in turn causes ulcers or heartburn. We experience fear, which accelerates our heartbeat. We have dreams affiliated with the good or bad conditions of the cells in the various organs and established feelings. We have hopes and goals. We have eyes that can communicate to our brain to be either pleased or upset at what is going on. We have intuitions and visions, dignity and generosity, etc. This makes it easier to understand that one ideology cannot satisfy that maturation process for all because of the differences established between one person and another across time. People can choose an ideology with high ambitions and, for example, with it the maintenance of a good economy, while others can or will choose to have less financial wealth, but more of the spirit and soul. Other people perhaps want less money but more morality and good consideration for the young and the old (not only materialistically, but in emotions and feelings as well).

Another example of choosing an ideology can be teaching children from a very early age to learn their future work without being supervised and not stimulated towards big gains. Maybe other people do not want ownership of factories but rather for the factories to belong to all the people who work in them and their families. Some may prefer private enterprise, while others prefer state ownership in a noncompetitive world. Others prefer the teaching of proper religion to their children in the schools. Some groups of people will take things as they are as long as they are guaranteed a job and provided with the principal necessities. Others want an ideology that gives in return that central natural line of life—dignity, equality on all fronts, full dedication to religion, protection of animals and forests, a sane mentality nor any aggressiveness, and working conditions without harassment, without supervision, content with minimum pay and minimum work.

In this world of ours we can accommodate seven forms of different ideologies, which as I said, can well satisfy all mankind at present. An ideology has good and bad side effects, because we are not living in a magical world. We must exercise dignity and determination, along with understanding, because nothing comes easily. It is not always sunny; we need clouds with rain. Our world is a creation in matter in an immense space, as we see, and we cannot always have solid foundations and an easy exit. We can use what we have as our Mother Nature accomplished in this living world and use it in the best possible way. Unfortunately, habits, misunderstandings, etc., lead our minds astray, to let others do the difficult jobs while we gain the best out of the results. There is no necessity to change if people are all happy to pursue any present ideology, but this is doubtful, because we have not yet reached that final ideological process of understanding, and certainly this causes pains and problems and people's unhappiness and revolts.

A competitive ideology promoting free enterprise can be good for the economy, but as people are too busy thinking about becoming rich and comfortable and paying for luxuries and habits to have time to spend with their children, so love and affection will lose here. Husbands and wives will have other things on their minds and cannot be romantic. A very busy materialistic life does not leave much time for friendship, love, dignity, respect, religion, etc. As a result many divorces follow, and prostitution rises, along with delinquency, drug use, and pushers trying to make easy money. Stiff competition in the workplace plus hard work can lead to depression, a rising blood pressure, and problems developing within the body and mind. On the other hand, with the increase of ailments, the demand for doctors, lawyers, police, medical specialists, pharmacists, and chemists increases. This in turn gives good possibilities of higher ambitions and goals. In a competitive world not everybody is strong and intelligent to the same degree, and this leaves an opening for the strong to become famous or rich or simply live comfortably, with a minimal amount of effort. A competitive human world can be interesting for those who enjoy competition, because everything is a gamble or game. If one becomes a winner, he can manipulate others to work for him, become rich, and live in comfort. Maybe your son or daughter was born strong and intelligent. This paves the way to future good gains to pay for comforts and for him or her to possess all the good things of life. Remembering that our Mother Nature gives chances for changes and it is understandable that it gives only basic necessities. If we want an excess of riches, comforts, etc., then we must use a competitive ideology, which in turn gives losers and winners as the "haves" and the "have-nots."

With religion, ideologies, etc., psychiatry made its appearance on the scene of human civilisation, but to be successful at this line or category of work (psychiatry) one must have people with mental ailments. Another ambition of life may be to become a supervisor or manager. One must be psychologically impressed as being better and above other people as an individual and in other terms to make more money, but along with this, it has to be an imperfect ideological world that gives poverty to many but is very beneficial to the few who are intelligent and strong.

Question: Do we have to then accept the world as it is, with wars, hardship, competition, and disease? No one can change the world completely, but we can remodernize our ways and make things more humane. To remodernize our ideologies and make them acceptable to all, it is better to lower our materialistic ambitions and behaviour of superiority and focus much more on a central natural law, as well as the acceptance of the principal good religions of this world. The decrease and increase of disease, bloodshed, hardship, poverty, delinquency, etc., depend on the ambitions this human race has and the attachment and obedience to the well-balanced principal world religions as well as physical laws. How diseases or ailments happen to be in our world is not much of a mystery. It is good to know that besides the diseases caused by bacteria, the rest are almost all man-made. How has man created depression, for example? Failure to reason properly is one cause; other examples are constant criticism or intimidation by a superior or coworker, being in the presence of bigoted people; an employee being under too close supervision by his superior or, as a superior, seeing others who are better paid and receive more respect. This creates a contrast between the inferior possibility of the worker and the higher possibility of the one who is the boss, which leads to tension and depression. Also, there is failure to obtain a desired way of life. Depression opens the door for other ailments, such as other mental problems, high blood pressure, delinquency, and suicide. With this domino effect, (one problem leads to another), then doctors, drugstores, pharmacists, psychiatrists, police, lawyers, judges, etc., become essential, and with this rises a different life from those of agriculture and industry. Ailments from the parents can be transmitted to offspring, causing the molecular system of the developing baby to fail to establish a proper balance, and can generate a body with problems causing the person to be less intelligent or less strong. Later in life this person eventually is open to failures such as prostitution, homosexuality, confusion of feelings, which in turn can lead to a limited vision of this created world, and from here to the nonacceptance of the normal things of this world, creating in turn that recycling of ailments that makes nature succeed in its raw accomplishments, as humanity always has problems.

Today as far as the natural chances for a competitive world are concerned, things are normal at this stage of time. But this competitive type of ideology has its side effects, because not everyone can be rich and a winner, but there are losers who have to work to support others in order to succeed in their ambitions. In fact, people with ailments give rise to the professions of doctors, psychiatrists, psychologists, and pharmacists, which gives employment to many workers. Delinquents keep police forces in good number, as well as lawyers and newspapers. The losers and the poor have to work hard for their employers, and this sustains the category of the rich or the possibilities of becoming rich. The wealthy people in turn provide, or at least should provide, work and basic necessities to the workers. The list goes on. As I said, a competitive world is a game to gamble, and the results are winners and losers. Anyone who finds this type of ideology attractive and adding to his comfort can well join in.

In this world of differences, it is natural to have people with different feelings and interests in life, and from here the birth of other forms of a "controlled" ideology, for example, and from here we have more employment for the workers, narrowing the gap between rich and poor. The bad effect of this form of ideology can be a stagnation of people's interests, money, comforts, activities, and luxuries. The good effects are that they can eliminate or cut short the number of people with ailments such as depression and many other problems related to the mind, such as aggressive feelings, high emotions, and delinquency. With this *calming down* because of less competition, people can become relaxed in their activities, and so their nerves relax. When an ideology is agreed upon, directions, for example, and gains, suitable religion, morality, developments, agriculture, etc., can be discussed as far as limits and extensions can go.

5

Ideas, mood swings, behaviour, and attachment to material goods can change from time to time in people. When that occurs, they can move quickly from one type of ideology to the other that suits their ambitions. No more revolutions or waiting for the ideology to accommodate their ways, they just have to cross the borders or boundaries. What is needed is a train ticket and a short time to get out.

It is interesting how our human body develops in its very first stage. We can see under a microscope that the various molecules travel the body and eventually they stop permanently in different places. Certain molecules form the heart, others form the lungs, others the brain. This world of nations is a body. People should be able to move accordingly to places where they fit in their ideology and in their destiny. As in the human body, in its forming stages (and after) group molecules move to their predetermined spots. So too, people of varying races and colours can move only to determined nations and continents. We must bear in mind that differences in our world are important for our physical and ideological life. This is why we have so many elements and ideas. To have a human world always under one ideology can be suicidal, but as we need a competitive approach for human automotion, then it is good to pull towards only one global ideology as long as in the end it will be divided again into seven main ideologies.

Besides the competitive and noncompetitive ideologies, we can make up a third one that can be a fusion between an ideology and a religion. With this type of ideology we can obtain an "Automatic Plan" in which harsh methods of law are not required much because religion can call a sin an infraction and therefore people in fear of sins will not reach their evil goals. The wealthy people will be much more considerate towards the poor. In general, people will be more considerate towards the people with problems and pains. Love and friendship will rise considerably and people become more respectful and happy with less delinquency, fewer divorces, abortions, etc. The side effect of this ideology is that because a religion should remain only at a spiritual level and not meddle with ideologies, many people will be tempted to gain more money and comforts than allowed and want nobody to stop them. This can create an attrition and apathy towards the religion and the church. Ideologies are the changeable parts on our human scene, and there are not firm limits and boundaries. Therefore, it leaves an open field for people to pursue "illicit" ways, and to curb this we have the laws of the ideology to set limits and the extension of any work, gains, emotions, and so on. Ideologies can be changed to accommodate our different goals, imaginations, thoughts, fairy tales, gains, emotions, feelings, desires, considerations, understanding, respect, etc., but our physical body remains unchangeable. In fact, our human body always develops the same structures: head, legs, heart, eyes, nose, and so on. But if the ideological system can indeed change, we must set limits and rules *not* to damage, while pursuing any nature of ambitions, our body systems as well as our emotions and a good religion and the belief in an afterlife. We also have to be aware not to spoil, harass, or damage the physical and ideological parts of other people or nations. In this world we have things that are irreversible. We get old; we die. Some physical and ideological ailments are irreversible as well. We need to counterbalance with a very acceptable religion with a promise of an afterlife and the touch of God at the end. Ancient Egypt believed in an afterlife, much, much before us. The truth is there are very good reasons to do so.

6

We have just mentioned a few ideas relating to three out of the seven ideologies I have mentioned. Now, on with the fourth. Always remember that our Mother Nature does not give a direct ideology, as in the case of our physical body. Our physical body is a "fixed" term; our ideologies are mobile. Fixed-term as it is, our human body is a mechanism, while the ideas give us the possibilities to differentiate from the fixed term of conditions. By changing our body structure and cells, we will certainly not survive. Changing ideas we will survive, providing the changes will not affect physical body limits. Having a sane ideology leads to succeeding in ideas and body possibilities and by adding a good religion we can obtain a very basic and complete existence.

An ideology is an *indirect* product of Mother Nature, and in order to keep it, we must work for it. It keeps us busy, but it also causes problems if we let things go too far, especially with the younger generation. Our human body is an accommodation of cells that survived across millions of years in the forests. For thousands and thousands of years our body cells accommodated an existence without work or obedience to any laws or religions. Emotions and feelings were not yet on the scene. There were no schools and mechanization. All that comes to an end in the last six thousand years, and that is certainly an impact for our mind and body, from the old system of the forests to the new in the last few millennia. From generation to generation the genes engineered a body conserved and formed back across time to our present days. Genes, D.N.A., chromosomes, etc., find the present world not so familiar as the one that has survived millions of years in the forests, and that creates problems.

On the ideological side, there is always a collision course between the younger generation and their parents. The parents have accommodated to the actual situation, but the offspring have not. To instruct the young creates arguments, gossip, and sometimes suicide (for the young as well the adults). It really should not reach such an extreme as suicide, because our Mother Nature, in order to give us a mobile existence (at least on the ideological side), cannot produce a fixed ideology as to the case of our body organs, which grow automatically, but we must instruct each other. This teaching is the good part of our existence, because by talking, changing ideas and images, etc., we exercise the feelings of ambition and experience new things, that is, *live*. Logically then, that our offspring must learn from the adults in order to accommodate to the existing era and accomplish a progressive and suitable ideology and religion. Many times people from teens to adulthood find it too difficult to accept existing ideologies because they aren't directly from Mother Nature or scientifically calculated, but created mostly by feelings of people. It is difficult to accommodate adults/young people to any form of ideology, because nothing is magic, plus when an astray mentality, feeling, etc., has been established as the "being" of humanity from one generation to the other it will be difficult to change. Many, many people have many question marks in regard to ideas and behaviour of people because we have a nonmagical world, with only natural physical laws, which it is difficult to obey and carry on.

Another thing that people find difficult is following the

only one ideology in the nation, because we are born to suit seven ideological forms. This creates dissatisfaction in the feeling and emotional system, leading to delinquency and revolutions, etc. Most of the time these occur because people see injustice or, by using injustice for profit, create situations where other innocent people will have to suffer the consequences. Sociologists, psychologists, and psychiatrists have always attempted to accommodate people to the acceptance of the ideology present in the nation without seeming to be aware that we were born with mind and body possibilities different from one another. This has always brought confusion and bloodshed in our human history.

It would be easy and good if our Mother Nature could have brought to the human race ideas and ideologies in the same way she brought the rain and sunlight and air. This is not possible, so we must do something on our own in order to move away from being or looking like just mechanisms. If we must work to find and adopt new ideologies, it is obvious that nothing is easy and nothing comes by magic. To believe requires our human mind to work hard, and this has always been a problem for the learning process of our generations. To have beliefs and attachments to religions also has been difficult.

In the forests thousands of years ago our ancestors did not have to work and think. Life was easy in this respect. With the beginning of civilisation and the first use of "intelligence," our ancestors began to understand and feel the benefit of comforts and the pains of working for them. Slavery followed. The invention of the wheel was to ease the pain of work. Our mechanization today is also to eliminate hard work. If our body cells had accommodated organs adaptable to the forest era millions of years ago, when no hard work or thoughts were needed, then it is logical that adults and the younger generation feel discomfort in working and obeying the teachings of religion and ideas. Such discomfort of work has created problems in our past civilisation because the temptation to use others to do one's own share of work always has been great. Religion had its problems, too. When the mind has to work for things, then the idea of believing in a powerful and miraculous God can become problematic and so a religion can be ridiculed and discarded. Another difficulty lies in that our human race has "standardized" more or less a course created in the forest across millions of years with no work or specific ideas and thoughts. So, according to this created situation, which nature has caused to progress in us, things become difficult when our body is forced to do work but become enjoyable when we have "plenty" without much work. This is one of the reasons mechanization was created. Our young generation finds it difficult to work, go to school, and listen to proper religion, and they must be constantly led in a good direction if we want to keep our civilisation acceptable and sane. If the younger generation is abandoned to their own intentions, for example, they could or will easily fall prey to short-term and shallow love affairs, noncaring in many directions, the parents always having to carry the responsibilities. The younger person could enter a career just for quick/easy money or have big profits with less effort to work. Delinquency could be seen as a means to easy money, as well as exploitation of prostitution, gambling, drug pushing, or simply becoming rich at the expense of others whenever the opportunity comes in one's direction.

As our Mother Nature is not involved directly on the ideological side, then we must understand that good teaching of the young (and adults) is the key factor in a good civilisation. Failure to teach proper religion and ideologies can turn our civilisation's development back in time or cause more of us to have a low mentality, being mostly inclined to do things unintelligently.

As we have a nonmagical world, just a natural physical mechanism, then it is understandable that, for example, in order to develop muscle cells we have to continually move. This gives us the possibility of becoming successful, and in turn those muscle cells can be genetically inherited by future offspring so they can move and work as well. This knowledge of how the muscle cells develop should keep our civilisation busy working and functioning. Supervision to urge us to work should not be necessary, because work must become an acceptable factor. A good ideology must keep the distance from those who want to gain too much with less effort. Otherwise "the spirit" of work in the people who have a good natural balance will suffer or die, and what also will die is the proper movement of civilisation forward. A high competitive ideology produces much work and many rewards, ways to live without work and a good comfortable life for many. With this we have a good raw natural condition that suits our contemporary times, but not good for a high, advanced civilisation.

7

We mentioned the ambitious and the nonambitious ideologies and the fusion of an ideology and religion, and these are the three primary ideologies. Four more ideological combinations can be added to the three primary ones to make the total seven. A combined ideology can serve two categories of people with different mind and body possibilities, serving the same ideology. For example, many people, despite their good intentions, fail to accomplish any goals. With this, their happiness becomes low and frustrations enter their living scene. They have the feeling of living in an unfair world because none of their high dreams come true and the spirit to work begins to die and the impression of a null and empty world begins to appear in their minds. They see around them many people who have succeeded in their goals and the feeling of inferiority begins to grow stronger and stronger and this world does not seem to be good enough to live in anymore. Frustration is the daily diet for these people, and the temptation for illicit profit enters their dreams. It is here that the ambitious ideology where they belong must begin or come in aid of these people, offering help to refresh them or to enter a career that can be rewarding in the end. This crossbreed between high ambitions and a "controlled ideology" will serve the purposes of raising up those who fail in their goals so they can walk strong again and can choose the ambitious world or remain in moderation.

On the other hand, the nonambitious ideology, as it is formed mostly by "moderate people," who do not have an excess of physical and mental strength, can be adopted in terms of comforts, work, goods, housing, caring of body needs, and good care of offspring. These people with low esteem have to be helped by the nonambitious ideology where they belong to become more ambitious up to that acceptable normal level agreed upon.

8

A fifth combined ideology could fit other groups of people where our Mother Nature has not yet succeeded in completing the maturation process. Our civilisation is a slow, natural process of many people. Through misunderstandings, many feel apathy towards many of the possible ways nature offers to condition a moderate way of life. As a result, these people barely grow in a sane emotional and spiritual and ideological process. Our Mother Nature is a workable solution, but not magic. To follow natural conditioning can cause the lack of patience of many, and from there people begin to rebel or feel prisoners of our daily duties and this leads to "imbalance" in mental, spiritual, and physical well-being. Their growth towards proper ways is then disturbed, and incorrect mentalities develop. Their deviations also affect "normal" people where nature has succeeded in maturing them in an acceptable and proper way. Physical and emotional ailments follow. People in this condition will not fit any normal primary ideology, because the natural offered conditions are not acceptable to their mind or spirit and they find it hard to be conditioned by human laws because of the deviant nature they have. These people could be corrected in proper ways, but the physical and emotional are mostly conditioned wrong and therefore many methods will fail to balance them. People who like unfair hard methods, such as high ambitions, et cetera, should only meddle with the same type of people in other nations, ones who would also be happy to use such methods and ideology. The problem with these people is not only material or spiritual ambitions but rather destructiveness and carelessness, with a lack of consideration. Therefore madness can only be satisfied or counterbalanced by other madness. Intelligence in our created world does not satisfy every human being. This can lead people to carry on in a very bored and frustrated way if a good deal of patience is not shown. It is enough to look at certain natural factors that make our world and have a quick feeling that something should or could have been better. For example, we have to die and any intelligent explanation cannot really satisfy us in a perfect way. We get old and with this comes the end of our dreams. There the explanation seems to be of no value as well.

Carnivorous animals eat other animals alive, and that does not satisfy us. Mostly the reasons and intelligence involved do not matter. We can explain that in reality animals will have to eat each other in order to limit their number or, as our Mother Nature cannot produce muscle cells unless we move continually, so enemies are needed in order to produce those factors that sustain life in awareness, controlled, to develop muscle cells, food, etc. In fact, animals could fail to move if they had no enemies because muscle cells could not have been developed, leading to muscular atrophy and the extinction of all species. This is an intelligent explanation, but again our minds raise the question of could there have been better ways than this?

9

Intelligence in our world seems to be low in its created ways, and we must fortify it with the conviction that our created world is beautiful, just, and full of wonders and there is the final welcome in Heaven one day. A sixth ideology can be then based only on a good religious way of life with just a little touch-up of the ideological laws. A well-balanced spirit with religious teachings can set automatic ideas of the acceptance of proper laws, work, obedience, love, considerations, and understandings without high ambitions or by the police state. Other people cannot agree to fortify intelligence in our world because the said religion and intelligence do not protect us against old age and death, and then they really do not feel like siding with our Creator and religions as a protest from a vengeful spirit. These people can group to form or choose the "dark" colour, meaning atheism (and with it the seventh ideology), if that is what makes them satisfied, their spirit less rebellious and sad.

The beauty of things can be noticed because there are other things with less beauty. Ideologies are infallible laws; we are not. This leaves an opportunity for people to fail, so many will have to pursue a bad man-made ideology so that even if things are of a lesser beauty and not perfect they will feel comfortable with what they have, as long as this makes them proud to go on and live. Logically, if our world was all a perfect level, rainwaters could not reach the seas and create swamps and we would be inundated with problems. Not all people are beautiful but, as an example, the ugly ones find their mates and are happy and proud. That also is the destiny of ideologies, natural or man-made, all of which have their portion of followers.

One of the principles of our human race to understand is that there is a good possibility that life will not end here on earth. There are reasons far ahead of our contemporary time that are not yet accessible. Our material body dies on earth, because the earth is subject to changes, but our Spirit (soul) belongs to eternity and will continue to live forever. It is then hazardous to have an ideology in complete absence of religion, as religion deals with the spirit of our body and, according to natural information, also deals with the real Spirit (soul). At the present time many people lack the advanced knowledge and this prevents them from believing in any spirit. It is much better if the nonbelieving group form their own ideology, separate from the believers. This prevents the believers from becoming spoiled or robbed of their realities. After all, if people do not believe in anything, there is no purpose in having religion. Besides, it is difficult to get attached in any way, even if forced to do so. They at least live in their ideological comfort in this life while other people can worship without being bothered and will have more chance for success with Spiritual possibilities. Our Mother Nature in her created ways can only offer conditions, these conditions are not magic, and so our Spirit (soul) cannot come unless we activate it according to that good natural given process in religion.

Our Mother Nature offers chances for ideologies, muscles, cells, feelings, love, mechanization, schools, and changes. It is up to our good will and intelligence to make things succeed. Religion offers salvation from that natural plan or possibility, and it is also up to us. If our world offers only possibilities we must not expect magical things, and so we must choose according to those possibilities we have been given. Failure to use natural, proper conditionings can cause any kind of problems. It is good to understand that our Creator could not have options if He had to create this world or not. An immortal Creator can have one side only, and the omnipresence of His existence is the cause for Creation as the world and what we see in the universe. It is like saying plants begin to flower when they feel the presence of spring. We humans on our planet created mechanization, housing, roads. God created the earth and the stars. If humans were not present we would not have roads, automobiles, houses. The presence or the existence of things created other things. For example, the existence of water, oxygen, and sunlight created fish, mammals, and plants. Pains, feelings, frustrations created intelligence and reasoning and from here to ideologies and inventions and all of the comforts that we have. The presence of religion and a good understanding of them and proper application can very well create our Spirit (soul) for an afterlife. Eternity has differences. After we die, our differences (water, calcium, feelings, thought) will remain, one way or another, immortal. Personally, I do not see what could be wrong with whatever form our soul would be with us forever. After all, it does not matter if things remain with us or not; it will continue to be immortal anyway. It is also very useful to know that if everything is immortal, then death cannot be forever either. In fact, when our human race will be extinct from earth, then it is enough to realize that the same process that created the sun and the planets will be repeated again and then, one day, our human civilisation will rise again.

For us everything seems to be confused or at random. But looking at our stars and our natural system, we see everything is calculated in every infinitesimal detail. Nothing is chancy; nothing dies; everything is eternal. Things that die will be turned and live again, if the process for rebirth will be given another chance. Mankind one day will disappear; so will our planet and the stars. Everything will return again as we see it now. Another chance, here or there. An unending circle.

The presence of a God can be the reason for what is created in the universe. A presence of an Entity does not necessarily mean the Entity creates things at will, but rather it is a "moving" process. As we are the product of the presence of a God created with the possibilities we see, then we cannot expect Mother Nature to do a magical process for us, but we all possess a form of intelligence that must help find the way out when we are in bad situations. The success of our human race depends on how much we are able to make our physical/ideological/spiritual possibilities work. We certainly cannot wait and cry when things become difficult, painful, or out of reach, waiting for God to do our job, because to keep our physical body and mind we must move.

As we enter the third millennium we must begin to trust our human race as an adult civilisation and not as children as in the past, so we can do things independently without much control and supervision. Otherwise we undermine the intelligence of each individual. The young can learn at schools and work at any level on their own, and that will certainly increase mankind's intelligence and self-esteem and with this our responsibility as an adult civilisation. It is good to know that wars, high ambitions, people with evil ideas, etc., aren't that good part of our given nature but a sign of immaturity. For example, if a plant grows weak and yellow under the shadow of big trees, that is not a direct natural process, but an indirect occurrence. A leader of state, a supervisor, and many ideologists can put their personal inclinations forward according to what their minds visualize, but it cannot be aligned to fit the multitude of people. If not, that can cause psychological and physical ailments. Therefore, individuals must choose what fits their possibilities according to normal and natural ways and from there be happy to carry on in their just responsibilities. Animals, for example, as they have just raw natural mechanical intelligence cannot analyse or reason about the world we have and are less affected by psychological problems than we are, but if they are treated off-course from their natural everyday pattern that can deeply spoil their normal behaviour and from there lead to ailments and problems. Eventually, if people one day begin to work properly without a superior behind their shoulders this can eliminate that "analysis of reasons" that produces the feeling that one is inferior and less human than others.

This alone can show a marked advancement towards a good, solid civilisation with fewer psychological ailments. Naturally, there are seven ideologies to choose from, and of course people can choose to work with or without supervision. Handicapped people, very young people, and certain abnormal people will always need supervision.

It would be beautiful if we could apply only one form of ideology worldwide. But unfortunately, we cannot. For example, we have many kinds of flowers, other plants, animals, and even our body system made of many parts in order to accommodate a full mechanism and many feelings and emotions, so the realisation of only one ideology to fit all mankind is a dream not possible on our planet . . . maybe in Heaven one day; or we will have to wait many centuries until human emotions, etc., will all merge and things will be reasoned out and the pains of human attritions will be better supported. Keep in mind that in our world things will not always go from better to best, but also the other way, too. The problems at the present time, as well as in the past, are because our Mother Nature gave us a process that must involve us gradually growing physically and mentally, so many misunderstandings arose across history. However, many in those early days in history had a very good awareness and intelligence and came up with many beautiful solutions, but these were not really understood by the multitude because they (the multitude) were still in an immature process. Therefore, those few that had knowledge of this world or where nature had already succeeded in shaping their ideas in an acceptable central line were defeated with their good ideas, by the large majority of people with not yet developed natural central understanding. As a result, those few people turned to being "ideologically silent," fearing they would be misunderstood and punished. Some wrote things down, hoping to be understood at a later date or in the centuries to come. Records of basic central natural good religious and ideological lines can be found in the Koran, Confucianism, Buddhism, Hinduism, and the Holy Bible. I repeat, in these people our Mother Nature succeeded in that good maturation process towards intelligence and to analyse the reasons of why things have to go in certain ways and so from possibilities rose the foundations for the principal religion on our planet.

The way our Mother Nature raised us from the forest to what we are today was more or less similar to the process of building a house. We must begin from the foundation. It is absolutely impossible to install the roof before the foundation. Our human foundation began from a primitive form of intelligence, to go up to the roof one day. Our intelligence today still needs a lot of touch-up ideas, divisions, trusts, and styles. We have much immaturity in the world today. Merely saying that we should work without supervision alone will spark furor in many because of the thought that people will not perform work accordingly. This is one sign of a still immature civilisation that lacks knowledge on how to make things succeed.

As our Mother Nature uses the same process, but with a difference, we can take examples from other living things and apply them to us. If this process works for other living things, it must also work for us. The growing process, for example, of a plant depends on many elements. The presence of one element gives the possibility for another to interact, with the plant growing and producing flowers and fruit. In our ideological process an idea can never grow unless reinforced by another. Ideological success working without supervision can be accomplished by the presence of the knowledge that our muscle cells will not develop properly in our present life and be passed on to future offspring unless we work or move accordingly. A person who moves *not* according to a basic natural good process can develop an everyday form of laziness, and that can make existence boring. That is the price to pay. A reason why lions, tigers, etc., are carnivorous animals is to keep animals in continuous motion and exercising and so producing a standard of muscle cells (one of the reasons). Without carnivorous animals or enemies there could not be motion at all. All animals would have disappeared a long, long time ago because of lack of muscle cells and the inability to move.

Departing from this principle, we can now easily understand that by attempting to defy natural central geometrical conditions we will pay a higher price in pain. Wars, high ambitions, exploitation of prostitutes, ideological problems, drug addiction and drug smuggling, poverty, many physical and psychological ailments, conquests, and slavery are a part of a created situation because of the failure to understand the importance of work and proper obedience to natural laws. It would be wonderful to live without work for all humanity. Unfortunately, this is one of the processes that our Mother Nature can never give us. If we pull the rope in that direction it will not break, but will break us instead. Can, then, this human civilisation work without supervision one day? Well, now the answer seems to be yes, as the know-how is now there.

10

The success of the growth of plants (as well as ourselves) lies in the presence of the chemical elements. But physical elements alone are not enough if sunlight (religion, in human terms) or other weather conditions (ideologies in human terms) are not favourable. The nonphysical part is exploitation, thoughts of inferiority, a superiority complex, believing nature is unfair because we have to work, expectations of easy money and grandeur, etc., which can kill the spirit to work, leading to the "unnatural" and thus the bad. Human genes are able also to accommodate a physical life inclined differently from that created central nature, creating ideologically and physically unfair conditions. Unfair conditions can be wrong determinations, wrongly choosing ideologies and aspirations, being self-centred and insensitive to the pains of others, unhappy under any circumstances, hardhearted and ruthless, sadistic, destructive, and blind. Understandably, then, our ideological genetic engineering is not yet so perfect in a big portion of people, while it is perfect in others. There are three primary ideologies and four secondary ones, to fit the conditions of people in their own genetic and present conditions. Out of these seven main ideologies hundreds can be combined to fit any condition. An adult civilisation must choose a way to live and a way to die. Also, they must be aware of the consequences of the chosen ideology.

Our gradual, natural way of growth created misunderstandings at every level of our human history. One of these misunderstandings is a developed feeling of apathy towards work; this is for everyone. To overcome this incorrect level in the emotional factor, people can choose a high ambitious ideology so there can be chances to become rich and with this accomplishment avoid what most people fear, worries and work! Other people can choose a nonambitious ideology, with less chance to work hard, also less remunerative, but creating many, many chances to ease the pain of work. One of these chances could consist of choosing a partner to carry on double-duty work for a certain period of time; then the other will carry on with same, giving the chance to each one to alternate that motion to work every day and with this feel less weight, dependency, and responsibilities towards work.

Other ideologies can or could accommodate work for only part of the year, one day on and one day off, one week on and one week off, etc. We have to abandon the idea that maybe one day we can use mechanization to replace people's work, because this can be catastrophic as well, against the natural plan for human survival. A humanity without work can reach a "dead-end mind" with the feelings of a "null and void existence" that can degenerate, causing bad and evil habits to contribute to further deterioration of the human spirit.

The human spirit towards work can also die when all people have a strong desire to reach for the top at all costs. To desire things high, to reach them at all costs, creates the chance of becoming insensitive towards our fellowman and, from here, lack of consideration, no love, lack of affection, and less of an inclination towards familiar responsibility (less time to spend with their children and/or friends). This creates a lack or an ennui towards emotions, feelings, happiness, religions, etc., and from here the general consciences and feelings in many will taste metallic or like machinery.

We have a human body that can only accommodate normal things in terms of work, thoughts, emotions, religion, ideologies, etc. But when we are overworked and worry to excess about fulfilling ambitions higher than that standard natural line, then we begin to lose control of proper natural feelings. Emotions begin to gradually disappear and with them happiness, serenity, and trust. Eventually, a high ambitious ideology creates money, luxuries, comforts, good housing, etc. But in this world, as we know, we do not all have the same ideas and there are of course a multitude of people who see other values in life and would be happy to be less ambitious, with less money and fewer comforts and luxuries or worries, and with this give more and proper time to feelings, emotions, normality, morality, and serenity to grow normally and accordingly. In other words, people are happy to lose money and material comforts but gain serenity, as more time can be dedicated to the wife/husband/family/religion, etc.

Inevitably we have a created world that never stands still. To expect that ideologies, nations, people's behaviour, etc., always stay the same is a fantasy, not possible in our world. If we were machines, we would simply push a button on ourselves and things would go on the same forever. Human beings have many feelings, thoughts, emotions, and experiences and have families, so pains and sadness can enter our human scene. We approach old age, we have bad experiences during life, our strength can run low, and so the will to do big things fails and the ideology that we used to subscribe becomes unfit for us to continue to believe in. Experiences, while we are maturing, come and go. People change moods, ideas, etc., and with these other desires will be born during life. If desires, feelings, etc., then arise on the human scene, we must accommodate them in the various ideological sanctuaries.

It is understandable that differences are the backbone and beauty of our created world. To have a perfect world is not possible, because the elements in matter or differences have no magical possibilities. The perfect civilisation, then, does not lie in achieving immensity and high gains in all expected ways, but in accepting bad and good conditions in their various physical and ideological differences. The day when we learn to be content with being given natural conditions, then we will be more than perfect to live in our world. Laws in nature leave an open space to form our awareness and learning process, etc., that leaves those changes to develop a random human behaviour. Our ideological factor is then mobile, so we can build our perfections or "a feeling body" accommodated among the dangers and the indirect process of nature and from here build up our awareness, emotions, justice, etc. For example, if two sparrows want to build a nest there is no chance of their building it wherever they want because of the danger of enemies. Let's suppose now that danger of enemies is the imperfect part of life that prevents the two sparrows from building that nest at will. Now to find a perfection among imperfections, these two sparrows will have to find a safe accommodation, perhaps a narrow hole in a wall where birds of prey cannot enter. This way the two sparrows will learn how to care, how to be aware, love, etc.

We as humans also have to find an accommodation among many unpleasant things, but this does not necessarily mean giving up or letting things run loose or at random, but finding an ideological and physical fitness or sanctuary away from imperfections and danger. Imperfections or dangers can be exploiters, people who use religion for evil purposes, disbelievers, warmongers, gossipers, hard-hearted people, those with delusions of grandeur, etc. As fruitful and fruitless trees live in our same world but not in the same longitude and forests, so we humans can have ideologies to fit each condition as long as we are separate from people who do not agree with our style of life. In this world there is room for all . . . but separate. It is useless to plant wheat, corn, and carrots in a forest and then expect a good harvest. Give them separate chances, and they will grow (and how!). Give people proper ways accordingly, and we all will prosper. Give people an unstable ideology, and it opens the door to confusion, ailments, and madness.

As the development of our intelligence on our planet is a natural process, we experienced through history changes and attempts to change at every level. New ideas were almost always on a collision course with old ideas. As rejection of new ideas followed, our human race found itself in a circular pattern that has always produced wars, starvation, and disease.

The attempt by nature with people to enlighten this world is thousands of years old. We can find examples from books of all the principal religions of this world. Hinduism, Buddhism, the Koran, and the Bible are full of examples. From the Holy Bible—for example, the prophet Jeremiah, who lived in the seventh century B.C.—we can learn one of the early developments of nature towards that maturation of understanding. With an astonishing accuracy Jeremiah said: " . . . and just as I [God] took care of uproot, to pull down, to overthrow, to destroy and demolish then so I will take care to plant them and to build them up." In this old script of almost twenty-seven centuries ago, we can understand clearly that natural process was needed, with wars, revolutions, etc., that uprooted, overthrew, demolished, and destroyed for the purpose of producing the intelligence we have today. With all these wars, uprootings, destruction, and overthrowing, it seems to be a pessimistic way to see the world, but in reality it was a natural process that also produced our human ways to take care, to plant us properly (ideologically) and to build us up. Jeremiah continued: " . . . when the time comes people can no longer say, the parents ate the sour grapes, but the children got the sour taste." Again, the word "parents" is interpreted as referring to our ancestors, because they could not have full interpretations of our natural process, so they were forced to test immaturity and from here derived many pains, as we know from history (sour grapes). As our intelligence will mature properly one day, then eventually we will tell our children and future generations that existence without work or magical and easy things are not possible in our world. With this the children (future generations) will get the "sour taste."

"To die of our own sin," in terms of our contemporary times, means to disobey that good natural process and from there to experience physical and spiritual ailments and death.

Jeremiah continued: " . . . the time will come when I will make another covenant with the people . . . I will put the law within them and write it in their hearts . . ." This statement is logical. When a new understanding will be processed from us, then it will be like a new covenant: "The new covenant will not be as the old," when nature across all those wars and diseases, etc., seemed to be as "a very drunk, confused and bad husband." With the new covenant and from our final understanding we can help to grow our proper feelings, emotions, loves, and ideas and from here have a good heart with its well-balanced natural laws. Jeremiah warned: " . . . as long as the natural order lasts we always remain in this planet as people and nations." Logically, if we try to run astray from Mother Nature or her laws we will be finished as a people and nation, etc., because everything will end in confusion and destruction. Also in the Bible is the foretelling of the advent of a star (Christ) and with it the birth of another necessary religion (Christianity) to be aligned among the great ones. Another "star" among the many in the firmament.

Our human intelligence is, then, a natural process that advances according to various experiences as can be obtained through contact, feelings, and necessity. People then, according to their experiences, came up with ideas to fit their conditions. Immaturity in the field of understanding our world created a human behaviour always in contrast. Ideas, laws, and conquests were attractive to some, but not to all. Our Mother Nature can only offer a standard central line, and any wrong that we do creates consequences or side effects. Our human genes can accommodate strength and resistance according to things with which we have always been in contact, things we have always felt, etc., across millennia from one generation to the other. Any changes in these factors become an unknown event for our body cells, which can be confused from their normal pattern. As we do not have a magical world, then geometry and gene history is what shaped, conditioned, and rebuilt our human existence. If, then, two plus two makes four, that is what it is and pretending to make seven, for example, out of it is like searching for things that do not exist in our world. In other words, it creates problems.

Our Mother Nature, then, has her boundaries or the drawing line in our physical structures. Conditions contrary to that blueprint can lead to physical ailments. For example, in the natural world of the forests where man lived for millions of years any change in weather conditions always was slow, as we also see today, but with the invention of fire, man came in contact with heat more often than in the offered natural conditions of the forests. That created perhaps the first form of arthritis, symptoms of colds, etc. There were the winters and hot summers, and with them our human body has learned changes in weather conditions, of course, but we were not so familiar with quick changes because this was not the case of the natural conditions always offered in the forests.

Returning now to ideologies, with the advent of any new ideology or idea, people do not always feel sympathy for it because of the unfamiliarity. As a result, many of the old and wrong ideas will persist, because people's habits, emotions, bad or good, etc., are attached to it, making the individual in need of either bad or good things because that is what formed body and mind.

11

We have a known history of six thousand years. In these years we have accommodated our bodies and minds to a pattern we know, with wars, conquests, dictatorships, high ambitions, superiority complexes, revolutions, selfishness, carelessness, atheism, blasphemy, insensitivity, and blindness in many ways towards the suffering of others—molestations, murders, and so on. Now let's suppose that we introduce an ideology that could change all those unfortunate human events. Would the people be interested in it? Maybe not, because our genes have partly accommodated human ideas and structures to be along that central line drawn by human feelings in contrast from proper nature. The drawing line of nature is a moderately ambitious existence with ideologies and the presence of one of the seven good religions. Anything away from that is wrong. Our genes, by human default, have partly accommodated our body's molecular and cellular structure, as well as an ideological feeling, and be satisfied in the wrong condition, then a moderately ambitious world and proper religion can be very out of touch even if this is a normal, natural, "given" condition. In good nature it is necessary to have seven ideologies. Some ideology can accommodate multitudes who have physical and ideological genetic malfunctioning. Good ideology can fit the conditions of those where nature succeeded in building them well. Once we deviate from that naturally drawn line, ailments will enter our human scene, also delinquency, child abuse, murder, rape, war, hate, and apathy for nature and the created world. A new thought can progress to gradually become an ideology. This can be wonderful, but the genetic code of mankind is too vague to feel any emotion towards it. If the ideology is proven correct and beneficial, it has to be accepted gradually and by the goodwill of people until it one day will synchronize with our genes and we begin to depend on it as our goal and our everyday well-balanced aspiration. In a changing world such as the one we have we must not put aside a coming "baby" ideology just because we do not know or feel much towards it. To put aside a new ideology means that we will remain in that painful circle we have known since the dawn of history. All the principal religions of this world had their prophets. Muhammad, Confucius, Christ, etc., did not have an easy life in the attempt to show us the real point where we should stand. Christ, for example, was sought out by the soldiers when he was a baby or a "coming new idea to live" and was crucified when the new ideas were known. This shows us very clearly that when an ideology has stabilized its roots, then right or wrong, it would be difficult to change. Eventually when a wrong ideology has stabilized as "living spirit" it has to live, searching, by rejecting a true ideology, religion, and justice.

Homosexuality, with a double or Siamese spirit, is wrong, but as things can be accommodated in the wrong way, then the wrong becomes alive in the body and has to search for feelings gone astray to stay alive. Nature does not condition us to do wrong purposely, but our immaturity can lead to conditions where people can be used as a "feeding ground" for others to pursue goals of bad nature. Thieves, gangsters, etc., know of their wrong spirit, but again the spirit has to live in a way that makes them comfortable. From the above example we can also say that our ideological world process at the present time can be wrong but comfortable to many even if all kinds of ailments appear in the human structure.

At the present time we really do not know where to turn and not many examine or question how ailments enter our human scene. Our pains then become an "outside" factor, with the thought that ailments come and go independently. Sooner or later people will look deeper into the understanding of our problems and then perhaps all mankind will come closer to that central natural line and will remain that way for a long, long time. Even at the present time we cannot pretend to give an ideology to people that in theory sounds good and live with the illusion that everyone should be happy with it. By nature we cannot all have the same ideas, and neither is strength, physical or mental, the same for all mankind. For example, we can choose the best and most fertile soil, with it expecting that all plants can grow perfectly. It is not so if we plant all together in that good soil, grains, poplar trees, carrots, oak trees, etc. The one that can be really successful, despite the good soil, is the one that can benefit from more sunlight. The poplar trees and oak trees, for example, but certainly not carrots and grain. This tells us clearly that because there are differences in things, then a good soil is a principal necessity, of course, but not enough if sunlight is not accessible to all plants.

Comparing the above example to us, a highly ambitious ideology gives one the opportunity to become rich, but the people who benefit from it are those with enough strength to support competition. Mother Nature certainly cannot offer a world with the entire humanity looking for big gains or big things; there would not be enough world for everyone. By nature, then, many do not have an attraction or strength of mind to be enthusiastic towards any high gains and high places. Multitudes of people do not focus much on richness and high gains because the natural world can offer other values in life such as family devotion, morality, devotion to the church, more leisure time, respect, more consideration, and so on. These people are likely to choose the third mentioned primary ideology, which is a nonambitious ideology. Also, many people even by having enough strength at physical and mental levels to reach much above nonambitious ideology are happy with it because of less work and less money and fewer comforts but gaining on the ideological good value of life. Certainly it is not good to give a nonambitious ideology to people who are interested in money and high gain; they will feel like prisoners if they cannot reach for high goals. Moderate people also feel like fish out of water when they are offered all types of ambitions in which they are simply not interested.

We came a long way as a civilisation without really understanding where the limits are that Mother Nature offers and that an ideological moderation cannot be accepted by all and in a short while. By default genetically we are not all engineered towards that natural perfection, because no one knows the chances to follow it properly and develop it. With time and conditions towards a more sane and intelligent civilisation we can change these stray ideas and shape better feelings and emotions if we persist in the good ways from one generation to another. For a good human success we can only follow those existing natural conditions, because feelings and emotion, as well much of the physical part, cannot be chemically produced. As an example, what produced our intelligence was necessity, pains, joy, care, frustrations, fatigue, visions, carelessness, etc. All these are not physical matters. If emotion, love, consideration, feeling, and understanding could be produced by medicine or chemistry then maybe we would have a magical world, perhaps much the same as our own Creator. It is better, then, not to wait for magic or nature to give proper feelings and emotions to mankind, but we must use and practise what Mother Nature gives us as a plan and a system to fit individuals' own proper conditions. Failure of adaptation to proper nature leads people to lack of consideration and not caring for others, searching for opportunities favourable to them and eventually using anything to reach the conditions that only they like. The rest, willingly or unwillingly, will have to follow. We have a created world that will always move, if not in the correct direction in the wrong one. It is of no use to have an ideology that is only comfortable to a portion of people and have the rest just as onlookers, without deriving anything concrete from it. Our life is short, and making a revolution to change things will take too long and maybe it will become a dream that will never come true. By winning a revolution there will not be much to gain, because another multitude of people will not like what the propositions are when the revolution is over or won. The beauty of each thing, ideology, person, and religion lies in the eyes and minds of the sympathizers and not the whole mass of people. We have seven main ideologies and seven religions where each beholder can be happy.

To advance in our human civilisation the question of leadership must also be put into question. Ideas, laws, etc., have to be used according to the requirements of the various times, eras, conditions, and stabilization. It is not necessary, for example, to keep the scaffold there after the building has been completed. As nature uses the same laws, but with differences, in a growing process, then in our ideological process we cannot stop and always use the same ideas. In the construction of a building we do not stop at the first floor and pretend that everything is finished. So too with our human advancement we do not stop at the present era but continue in the changes according to "natural process." Leadership, for example, has been good in the past, as well as today, in order to move in some direction, right or wrong, and in that motivated people into understanding and corrections. Obviously when corrections and understandings are reached (in the years to come), it is not intelligent to keep leadership any longer. Otherwise, it will look like a finishing building with the scaffold left there permanently. If we do not need leadership forever, what will we use instead? Well, as our human understanding grows and expands it can be understood easily that only one person as a leader of state cannot answer, feel, counsel, ensure, etc., the various demands of the multitudes. To answer all the questions, feelings, ideas, etc., of the people, leaders need a university education in the environment, psychology, theology, psychiatry, and science in the direction of physical matters, and so on.

In other words, then, as our civilisation grows into more understanding, we need a "team of leaders" of a state to satisfy any demand from the people. This is not to undervalue a leader of any state, but the reality is that we must have degrees in many fields and this takes longer than a lifetime. It is better to have a team of state leaders to simplify things, or the various feelings and demands from the people. For a suggestion we can say that a leader of a state can be advised by professionals on what to say, but to say things without proper feelings, intuition, understanding, and consideration is not advisable at all. In our past history leaders of tribes, states, and empires had only feelings and intelligence devoted to provocations and wars. Logically, by not having any pillars to erect the scaffold (people instructed towards psychology, the environment, understandings of ailments, good priests, etc.), we always fall, as in a domino effect, into the same mistakes of wars, hate, madness. From this natural point of view we are not yet finished towards that final understanding and cannot take the scaffold away yet but add much more knowledge to build other levels until the end of the building (civilisation in full). In the past, leaders of states seem to only have been interested in wars, conquests, and superiority.

12

In the working situation worldwide supervisors, foremen, etc., seem interested in superiority and production only. Production, in a moderate way, is “one” of the necessities of our human race. There are also other necessities to consider as state of mind of the various workers, physical and mental conditions, religious beliefs, what situation the various workers may be experiencing (bad or good), family and/or financial problems, age, aspirations, and treatment of others and vice versa. Frankly, not one of the leaders of this world has enough intelligence to deal with a multitude of problems, because to understand all would take a learning process that could last a lifetime. To take into account only a lot of work and production, forgetting other necessities in our human life, we actually deal with a primitive person whose lack of knowledge makes him or her a “metallic” being with no feelings, consideration, and emotions. As our Mother Nature could only progress us from a zero point of understanding, it is good to know that leaders of states, supervisors, generals, etc., were necessary. But as I said, we take away the scaffolding when the walls or the building is completed, because it is then able to support itself. When we were babies our mothers used to hold us when we were trying our first steps. Later we could stand on our own, so our mothers held us no more. Our mental process one day would allow us to move independently.

It is advisable that people study in school how to function without supervision, which they will need to know once they become adults. Having people always tell us what to do shows that we are not autonomous or do not understand things by ourselves but rather are mechanisms that are incomplete or have to push many buttons in order to go on. A human world that fails to accommodate a natural process can fall prey to depression, delinquency, and death of the human spirit in many departments, physical and mental ailments, and spiritual nonaccomplishment. We must keep our human minds always active; we need new growth. Eventually one day we will understand everything regarding ourselves and the planets, but this does not necessarily mean that there will be the end of human civilisation. In our world nothing is magic of course, but when a building (an ideology) is mastered and our human minds start to feel less room for advancement or begin to feel as if we are prisoners or uninterested, then we build another building (a new ideology) where it (the human mind) can move again to accommodate and extend according to the room it has been given. Ideological growth is a pastime and factor to live for our human spirit.

We have seven main ideologies and across centuries will follow-up with new ideas and perhaps new ailments, discoveries, etc. The ideology can be built or rebuilt to better accommodate feelings and motivations of the various groups of people of future times. Actually, our human minds have three ways to move. First is the time for the learning process. Second, we must keep for some time what has been mastered across its uses. Third, we must wait until the maturation process is over. When people become bored, then it is time to change again. Boredom is transmissible from one generation to another across genes, and after a period of a few centuries ailments related to the type of ideology (side effect) can rise. So it is opportune changes and with it exercise of part of the mind that has not been used for a long time. Idleness in our world cannot be applied for too long. Otherwise, it can turn our way (feelings and motions, etc.) towards stagnant or swamplike patterns. A circulatory or a ventilated process has to be applied to our ideology to give mankind ways to move avoiding stagnation (depression, boredom, ailments). In other words, when a person or group of people chooses a nonambitious ideology, for example, it gives rest to the ambitious process (which also exists in any mind), which in turn opens rooms to the application for it (the ambitious ideology) and with this the procedure for an ideological circulatory motion. Water, air, etc., always must move, and so must our human spirit.

Nature uses the same laws, but with differences. So differences must be applied to our ideological process as well. Our present ideological process is as old as the world; it has been kept for too long. Ventilation or change is necessary for our ideological factors. Our heart, for example, has its own method of circulating blood. Leaves on a tree have their own cycle and circulation. The wind moves the clouds and the waters, etc., and that is another type of circulation. Our heart, leaves, etc., need structures to make circulation possible. We need many ideologies in order to procure circulation in our mental process. A moderately nonambitious ideology with the implement of a good religion is the backbone and proper natural central line for mankind.

Unfortunately, things cannot stand forever on the same base in our world, because we always need a circulatory process. Therefore, the aforementioned moderately nonambitious ideology can only be the centre, having at its right (but separated) the highly ambitious ideology and on the left a corrective ideology with not much flexibility in its laws, in order to stabilize wrong human ways back to normality or according to good nature. So the ideological circulating system goes from the said centre up to the right and left and back to centre, and so on. Circulation, then, in our world is part of that natural structural process. People moving from one ideology to another are the parts or particles of circulation.

In our human social structure we have labourers and professionals. This is not a distinction or unfairness from Mother Nature, but a facilitation or "chain motion" to better facilitate human necessity. As I said before, religion is an aid to our human intelligence. In our working conditions a talented worker with a university education is nothing else but an aid to manual labourers. It will take a long time down the road to human civilisation in order for each of us to learn various departments and be a complete worker. According to our contemporary civilisation and natural process, the burden now must be divided among the people. A manual labourer then has a helper, a professional worker, whose guidance makes things easier. Unfortunately, we had and have our misunderstandings regarding our world today and professionals, who thought they were superior people, developed the idea that they should be better paid than the workers and from then got more respect from others. This developed our human conditions more or less in a form of discrimination and a form of slavery instead of that aid intended by nature.

Machinery also served as another form of aid in general, aid to hard labour and people with ailments or weak in the physical and ideological conditions. But again, unfortunately, it mostly has been used to make quick money and enable some to be "superior" among the same people and other nations. This is nothing else but another example of immaturity in our mental process. If we use machinery too much to make profits instead of facilitating or giving opportunities to rest from hard or fast work, then we attempt to break that logical natural process that is to keep us busy or move continually for reasons of muscular structures, awareness, and mental activation. In reality our world is not for big profit, nonsuperiority, etc., but a way to succeed and survive.

Eventually, speaking of ideology again, if there were only one central nonambitious ideology then that would be it. Unfortunately, we need motion.

It is understandable, too, that if animals could motivate themselves properly on their own without causing extended damage, then their enemies would not be necessary to keep them alert and active. We as human beings have the possibility of possessing intelligence and knowledge, so maybe we can break temporarily or disassemble raw natural conditions to build a smoother mechanism to serve a more humane society.

As this world has been created, then it really does not belong to anyone in particular. As long as a group of people remain sympathetic to following from their hearts an ideology that does not produce many side effects, is not too disastrous, and is still based on that good nature, then they have to ask no one for permission, only the government for assistance. For example, many people perhaps want to demonstrate that God is righteous and has created a world equal to all. According to natural conditions this seems not to be so, but we can use natural ideological material to build differences that can work for equality. For example, many can be fed up with people who have good jobs having better working conditions and remuneration than labourers. Either side one day can decide that as people are all the same, one cannot have a clean job for which he is overpaid while others have much harder, unclean jobs for which they are underpaid. So, in agreement with a group of people interested in more equality than we have today, they can decide for a common accord and, after thousands of years, follow a different pattern by instructing children from the beginning of their school life until maturity all the knowledge necessary to have an office job, to be able to work on the farm, to be labourers in factories, and to perform many other related jobs. In this case using a rotating manner, everyone will share work in perfect similarity. Naturally, this cannot apply to all, because some of the professional people will have to always keep the same category of work. This type of ideology can be applied only to people with a strong mental and physical condition and at least a good degree of love and consideration and very well educated and well advanced in our present human civilisation. It is necessary to be strong because our ideological and genetic inclination at the present time really does not support equality, because across past centuries people have always been motivated by superiority complexes, using people for personal advantage, using criticism in order to obtain personal gains (and friends), superiority, intimidation, slavery, and so on. It is then understandable that at the present time we, ideologically and genetically, are fit for inequality as people, as a nation, and as a world. Not many people, then, can pursue the abovementioned ideology, because they are genetically attached to being more satisfied in the old ways and so feel no emotion or sympathy for equality, because there will be nothing to criticise or curse and no chance for superiority complexes, etc. If we use more equality, then, our world or Creation and God will become more just. Besides, it is good to see and understand things in advance so we can prevent the future from following a dangerous course.

For example, let's analyse how we can prevent the future from falling into disgrace. As Mother Nature gives us a centre line, then a normal constant motion and use of strength are what really gives muscle cells the chance to develop in condition and synchronism with the offered natural world that we have.

What can happen, then, if people motivate themselves "under" that stabilized natural line? Their offspring in general can become handicapped for a normal motivation in our offered world and job conditions. Let's now suppose that the mind is too weak for work, improperly obeys, and advances poorly in school. What can the side effect be? Probably these people will have an ideological temptation to steal, be involved in illicit commerce, prostitution, etc., as easy ways out. Gangsterism can also be an attractive way of life, so with this attractive way of life one can be pleased to become a supervisor or reach any easy high place or live on taxpayers' funds or just live off somebody else.

The mechanism of our Mother Nature always attempts to give us a good central balance, but cannot succeed in this when we pull things too far astray. A female egg is tempered according to past descendants, but has to accommodate a newborn to the present era, so welcomes a male spermatozoon suitable for contemporary times. For natural corrections, not all offspring are clones of the parents, but can be projects or projected for futuristic rules. A battered mother, for example, bothered with hard work and unfair circumstances, hardens her physical and emotional condition and also her female eggs. These eggs will not have a good natural balance or centre, and of course the "egg chemistry" is not a normal one. Abnormality in the egg chemistry welcomes and interacts particular male spermatozoons to close or repair the fall across the offspring in the era of civilisation in which the baby will be born. Classic examples are that we had many good scientists and political figures in big ways. We had Confucius and Muhammad, the prophet. And in minor ways we had many other thousands across history in order to put back in place a human world that was going astray.

We must not always depend on the natural plan to be dragged away in place at any time we do things wrong, but we must learn to understand and follow accordingly. In this case, the innocent do not have to pay the price to get humanity back in place. To begin to prevent bad things we must rotate human behaviour in terms of ideas and work; otherwise we can lead to an imbalance in weak human ideas and muscular strengths, which in turn leads to laziness and then to apathy and lack of interest in that category where nature requires calm, steady work.

Most people in the world today have lost that happy/strong feeling through work in offices, fields, or factories, because their parents and ancestors perhaps had constant ideas of easy gains with little work or no work at all. This genetically diminished the human strength and pleasure in work. People are now in the situation where they have to "force themselves" to work, while in the normal way either strength or mental "will" should be generated spontaneously, with an easygoing happiness.

This tells us clearly that the way we act and behave has an effect on our lives as well as on the genes of our offspring. It is also useful and logical to understand in advance all the side effects of any chosen ideology. Understanding an ideology will likely allow one to know what to expect when a tree is planted. It can be a fruitful tree (agriculturally minded) or fruitless and used for lumbering (industry). Ideologies can change also in order to cut short certain people's bad habits, such as delinquency for easy gains, no inclination towards industry, agriculture, commerce, morality, religion, etc., as well to correct certain psychological ailments. An ideology that is very wrong cannot be pursued, because it is not a sign of good nature but a wishful "trajectory" by physical and mental problems of individuals. However, if people want to pursue wrong ways, and our actions of stopping them would prove more harmful and more painful than the ideology they are in, then we must not interfere. So the lesser of the two evils has to prevail as long as they do not bother people who feel comfortable with other forms of ideology.

Our world is not magic, and our survival depends on the limited possibilities of Mother Nature. It is not good, then, to ignore, avoid, curse, and criticise the natural laws and refer to them as stupid. For example, many people in this world think and have thought that religion is not an important factor of human life, preventing the good natural use of it. This is and was a sign of people's immaturity, because it prevents one of the many pillars for individuals to become a "complete being" as Mother Nature planned for us. It is wrong to take away or put aside many of the natural factors that shaped, kept, and stabilized genes towards life's success, as we see. The success of nature consists of information, needs, adaptation, etc., and as our earth geometry is a circle, then our body can only reproduce itself. Therefore, as long as that natural plan will last, our human cells will reproduce in a sane manner, as stated, producing our body structures.

To change things or feel apathy for our Mother Nature and her laws could possibly mean failure to mature into functioning creatures and generate confusion and ailments at physical, mental, and spiritual levels. The problems of today's world mostly consist of us wanting to insist on curing everything with medicines instead of giving nature the full chance given by Creation. We do not understand religion or ideologies and their applications, in a comprehensive manner at least. We overreact and overwork. We think too much, and everything seems to upset us. We have no attachments and understandings. We believe in nothing. We expect magic instead of a mechanism. Behaving in those ways and conditions creates ailments that we fight with medicines. If we join that natural process, maybe we do not have to use chemistry or medicines to be aligned with that central line offered by nature for our survival. Many times we understand that we are wrong, but we do nothing to change because we have created the feeling in us of comfort with these wrong ways. By doing wrong, we prevent proper stabilization in us with proper ideas and genetic information. In addition, we spoil other people's proper ways (if any) according to proper nature. When we do wrong things we also attempt to build a defence to keep our wrong, which makes things even worse. A synchronicity in all structural level is a full success for life. It is not good to have a perfect motor (physical body) in our automobile but not give interest to or not take into account the electrical or electronic parts (religion). It is also not good to give full recognition to the motor and electronic parts of our automobile but fail to consider the transmission and brakes (human behaviours related to the possibility of the body and mind).

Obviously, then, we must keep our Mother Nature in the ways it has been created and wants us to be, and with this keep the "factory" (good natural process) that produces our cells and from there our human body and the rest of the good condition as we see or would like to see in its enjoyable effects. What keeps body and spirit is air, water, soil, food, exercise, necessity, natural pains, religion, ideologies, amount of work, and thoughts balanced according to our possibility of withstanding with our human body and mind. All these factors must be evaluated one by one, and if we succeed in keeping proper ways, one day this human world can really be at its best, with fewer pains, ailments, and problems.

Pains in our world are a part of Mother Nature's system of pulling us towards that central line in order to give us an identity, reasonable feelings, a body structure, and a way of intelligence, defences, and consideration. If we did not have a central line, it would not be possible for us to be attached to any law, have any emotion and affection, and have reasons and a life-style to follow. Without a central line we could have been nothing else but something without identity. It does not necessarily mean that we must forcibly base things always at noon, because we have a circular flexible and space limits so we can physically grow and ideologically move. Not necessarily all the changes have to be for the better. Logically, everybody likes springtime, but unfortunately, the summer has to come in order to mature the fruits. We would like to see that our children have everything that they desire with no work at all, but in this we spoil in them awareness, consideration, understanding, etc.

13

At the beginning of time, when all substances were not in stars and galaxies but in some unknown form, these said substances had no geometry, there was no centre or any reaction or laws, and there were no magical factors either. If nothing was magic, then we must figure out that in order for us to live on this planet there surely must have been a long, laborious, and ingenious master plan. In reality, then, substances or matter do not possess pain and joy, but this can be accomplished between differences in life and the elements, which can form the soft and the hard as well as cold and hot and, in a more elaborate way, sour and sweet, and from here to the development of our feelings such as gentleness and toughness, love and hate, pain and joy, etc. Sweetness, gentleness, love, and joy are nothing else but the soft ways between hate, toughness, sourness, hardness, coldness. Example: for a comparison to soften our feelings we can look at the life-style of those animals in the forests and fields that really have no safety of any kind, we also notice the number of enemies they have. Their survival rate and condition are heartbreaking. We as human beings have many advantages over all creatures, and this makes us ideologically or psychologically able to support our pains. Space, for us, means emptiness, and this factor gives a chance for matter to identify itself, plus use this emptiness or space to make its own spirit in living things. For example, that geometry in the structure of living things could not have been possible if space were not there. Otherwise, our planet could not orbit around the sun. Our ideas, vision, colours, feelings, also spirit, could not have been present if that immense space that we see was not there to be used and copied.

Returning now to ideologies, it could be good to use one ideology for all, but the reality is that the human mind moves by itself in some direction, because during our lives and from one generation to the other we see and saw changes. As our human spirit then grows with determined changes, we have to welcome and accommodate those said ideological inclinations of people. If we do not, then people, especially the youngsters, will have a tendency as to become hypercritical, bored, and weak in the mind, looking to pursue strange ways and unfamiliar behaviours. It also has to be understood that our world has limits, so strange behaviours, hypercritics, etc., can come to the human scene because we expect things that this world or natural structures can never offer. In our world ideologies cannot move as a straight arrow to infinity, because they have as a base and foundation a natural mechanism made of circular geometry. As the situation is, nothing is immortal in our world, and ideologies also cannot have new parts forever and ever. But as we humans are not immortal either, gradually using our ideas, learnings, experiences, accomplishments, et cetera, that will keep us from childhood to old ages. But if we use our learning, experiences, et cetera, too fast, then one day during our life we will have nothing new—here is where our spirit will begin to die.

After we die, a newborn will come, so whatever they learn or see is fresh and new and that stimulates and attracts them to carry on. If people never died, then, our world would be limited. We could not experience new things and could not have that stimulus to carry on, causing our spirit to die. In some other examples, if our heart stops circulating blood, our human body cannot be supplied, plus the blood will coagulate and this stops our existence. If we will not recycle and circulate our ideologies we can reach a stagnation of the mind and our existence can seem bored, causing depression as well as the impression that we live in a world that does not have much to offer. In reality, it is not a sign of high intelligence for people to study things quickly and understand most of it at a very young age. Eventually, if we had a world where an infinite line could be used instead of a circular one with limits, then it would be smart for children to learn everything as fast as possible, but as our world has limits and the word intelligence has value only by following proper natural laws as we know them, then it becomes logical that our learnings, experiences, work, ideas, etc., have to be administered in that amount that enables us to be busy from adolescence to old age. Ideologies, learning processes, are, then, the life and pastime for our brains. This pastime becomes our stimuli not for our personal goals, but it has to be gradually administered to the individual progress from childhood to the adult stage and it is unsafe to use too much at one time. Certain ways of life, reality of life, stages of life, etc., can only be mastered gradually in children. Otherwise, by going too fast we will make natural chances fail to grow in unity in body and mind and then the ideological process or attachment that should last from childhood to old ages will fall short and from here the feelings can be shortened sooner or later in life and we will feel that this world really has too little to offer and our old age then will really become bored, frustrated, and empty.

It is also true that frustration, depression, and emptiness can rise through not obtaining things that this world can offer or by expecting too much more than natural conditions can offer or by unfair treatment from people and self-illusions. We must begin to look to our world as a "creation." A creation is nothing else but a product. As our intelligence is then a product, it must be used in relation to the body and ideas, phases (the growing process of an individual), necessity, and life span in order to keep us interested until a very old age. We can also understand that as our intelligence is a product and is to accommodate our life in this world, we must not get carried away so that one day we can explain the whole eternity. Everything is related to eternity, and so even our form of intelligence one day will go very far, but the fact remains that we have been created and assembled in matter and cannot become space or spirits or God, because to be like that we would need many different ways other than flesh and blood.

All religions in this world tell us of rewards in another world after we die. The possibility of another world is very real, because the "know-how" from God has made it accessible, but even living there, we in another world in Spiritual form and forever, does not necessarily mean that we will control eternity and be God, but we can live in God's kingdom, in other words, living in what God's possibilities could have created, a Heaven to welcome our terrestrial Spirit. We cannot fall into temptation, making another observation, and must not believe that we have been created for evil purposes against our advantages. If God is a Creator, this, frankly, means that eternity is the unchangeable factor and cannot be defeated and so no one could use it for personal evil and horrors. It took billions of years in order to raise life on our planet, and this tells us clearly that God could not create things at will and in a short period of time but had to give substances what they could bear in order to become a world. It means, then, that God was more or less a helper for substances in becoming what we are. This is also an encouragement for us to clear our human minds of some evil entity that could have created our lives for evil purposes. The success of God in creating us lay with putting substances in a place where they could develop as the world we see. Space, for example, is a place where substance can form galaxies, stars, planets, and a sun for the earth to orbit around in order to give chances for elements to interact and produce life as we all know it. If the matter we have on earth and in the stars was given another place instead of the immense space we see above, then logically we could not become the kind of life that we have become. If eternity, then, is really "in charge" of all, no God or evil force can create a kind of life totally away from reasons. Complete evil purposes would have not worked out.

Many times man has observed lions, tigers, etc., devouring other animals, and we reach the conclusion that perhaps we have a bad Creator. The fact is that these are the ways substance would live life as their possibilities and not exactly as a Creator wanted. In other words, then God can really have great love for us and the created world but had to give substances only those possible chances that were able to become life as we have it. If the temperature of our sun is millions of degrees Celsius, it does not mean some kind of Hell, but only the right chance that substance needs to become workable. Perhaps there are other methods to condition space substances into something other than what they are now, but then we could not certainly form life as we have it today. We can also easily understand that God is intelligent or possesses intelligence; otherwise He could not have known that space, time, conditions, etc., could have produced us one day.

God had ideas and with them came the genesis of our world. After that, all things are progressing automatically or in what we call an evolutionary manner. Obviously, God does not need to come onto our planet to cause our ways to progress because everything now belongs to that automatic natural plan until the end. Everything is a synchronized process and reason.

One of the reasons that could explain why stars are far from us is to keep that safe distance in order that, as there is intelligent life here, no one from elsewhere can meddle erroneously, especially when intelligent life is at the very beginning on our planet or in any planet and by not understanding proper ways could give ideology to spoil totally or badly their automatic or natural given conditions. In other words, let's suppose that the moon had intelligent life. Then it could happen that with these ideologies earthly people have, but never understood, then it could have created more attrition among them, triggering not only dangers for our nation and world but as well for the imagined intelligence supposedly on the moon. One day in the distant future when we have substantial knowledge of life on other planets and the stars we will leave things alone; that prevents our ways being spoiled as well as theirs. The fact that no one has really come to our planet from another world is mostly because there really is no need. Everything has and is taking care because we have a workable, complete natural plan. If nature fails to become as planned, this means that our planet is defective and as our substance or the planet is not magic, then in reality no one can really do much to save our ways. Our world is a process initiated billions of years ago. If we have wrong things in the natural plan, we certainly cannot correct it now, unless we go back the millions or billions of years ago to when things started to go wrong and fix it. Frankly, I believe that no one can do that. This can explain why despite our many pains and problems, no one came from the stars to save us from the problems. If things are correct, it means that the natural plan is in full control. If things are bad, it means that nature is not so fully in control because of human failures. We now cannot be carried away in the believing that our excessive pains and problems do exist because our planet came to be wrong, but because we fail to accommodate things properly. If a planet was created wrong, it would not have produced any form of life anyway. A wrong planet set its course or orbit either too far away or too close to the sun and with this prevented life from forming, avoiding with this further complication, horror, and pain for the creatures. A planet lacking living creatures means that it has not been well stabilized and complete in elements to originate nature for any living creature. Excess of certain elements or the lack of them is perhaps what keeps the planets distant or closer to the sun.

14

Ideology, religion, etc., in our world are as important as anything but came very slowly and with much contrariness and misunderstanding by people of all eras. The invention of the wheel by our ancestors created excitement because the wheel was a physical and visible thing. Things were and are not so interesting when they are unseen and need a great deal of understanding and courage to explain, especially when we speak of psychology, psychiatry, ideologies, and religion. Even today we still believe the proper spirit, psychology, religion, ideologies, and psychiatry are not important. There have been many martyrs in the past in the attempt to give us a world more humane and central. Christ, for example, was put on the cross because he was attempting to explain and promote the importance of the unseen part of our world. Evidently, then, our ancestors were not aware of the importance of human behaviour and did not understand that they should correct their bad acts and from here they went to being heartless, etc., and crucified Christ.

Prophets like Muhammad and Confucius did not come in close contact with the real explanation at a physical and ideological level but mostly spoke in a way to satisfy people and alleviate their pains through supernatural help and Heaven. That really caught the imagination of people in those early days, but as we all know, pain, disease, starvation, and wars remained and still remain today. To explain the realities involved in our world takes a very advanced civilisation. Today's civilisation is still not mature enough to understand the many natural concepts of our world and can become aggressive and nasty, cursing when someone attempts to put things into a proper, naturally assigned destiny. When certain habits and behaviour of people are conditioned and progress abnormally, that can become an acceptable way of life and can be profitable. A feeling of superiority and even hate and sadism can be satisfactory.

It is hazardous for anyone to deviate from nature in ways not intended by Creation. Once nature is deviated from, it will develop present and hereditary ailments. Here too when things are wrong, they continue to be wrong, but that does not necessarily mean that things can never be straightened out. Of course, if we are conditioned in the wrong way we still live and have interest in life, but it is painful and almost everything is irritating and engenders hatred. We have chances to build a good physical "human world" according to proper natural laws that produce the will to work, good acts, interest in loving life and all living creatures, plus the comfort of good religion/ideologies, for a normal spirit and human behaviour, etc. Would this be a much better civilisation with less pain and problems?

Life on our planet is more than just millions of years old. This very long time was needed to direct all living creatures in a way of order, self-automation, feelings, and intelligence. Changing or hoping to rebel against proper natural conditions is an unintelligent way, because we will be looking for something that is not there. Sometimes we can even succeed in changing natural proper conditions, but it always produces pain in all directions. To change natural, proper conditions we dismantle what this created world offered across millions of years. We have the right to move according to the natural elasticity that has been given us but keep our ways and acts away from spoiling those conditions that have raised us and will keep us, such as religions, ideologies, language, moralities, the Spirit (soul), emotions, feelings, laws, destinies of nations, leaders of nations and superiors, plants and animals, keeping water and air unpolluted, limits of mechanization, fertilizers and medicines, limits and uses of computers, extent and use with limits of television as well as radios and movies, the amount of work acceptable for each individual. All the aforementioned conditions are what form our spirit and the physical "being"—in other words, mankind. By keeping a standard concerning the aforementioned factors the future generations will have not much pain and not many problems. Acting contrary to proper nature, everything becomes a nightmare with not much order, not good feelings and emotions. It is like dying a little any time that our natural conditions are damaged. An automobile, for example, when we take away parts from the motor ceases to run properly or at all. To keep our human body and mind healthy we must keep those ways that nature can properly produce a human body, organs, and the spirit. Failure to keep nature on course will force us to search cures from medicine, surgery, and psychiatrists. Many times we find out that surgery and medicine will not always cure us and so psychiatry will not cure us either, and then human destiny many times is forced to a premature burial or to be interned in a mental hospital, many times for good.

One of the many concepts that we have today in the world is the misunderstanding of religion. Religion is the most important pillar for good stabilization of feelings and emotions, spirit and goodwill. By cutting the supply line of this natural condition of the application of a good religion, our spirit spoils itself and we become insensitive, with no love and consideration. Also, it raises the question of why we live in a world with so much pain, for which most of the time we do not have any answer. In order to get answers we need to study and understand natural plans, and from here come all the explanations that we want.

Mental immaturity can be as bad as a killer. Many people in this world think that religion is a thing of the past, an antique, a figment of people's imagination, and with this we really prevent nature from coming to us for our proper conditioning. If all were magical, then religion would not be needed or, perhaps, was only needed at one time, perhaps back in early ancient times, but even so we always have new generations to follow and they need teachings for their portion of balance, conditioning, shaping of their good feelings and emotions, as well as love and consideration. Without proper religion either soldiers or police will have to work very hard to keep people in certain order, to prevent them from stealing, killing, and destroying. Even so, people obey the law for fear of imprisonment and other consequences, but still their emotions, love, and good feelings simply are not there. Once we take away what produces proper feelings we become unconditioned and go astray. Going astray is as dangerous as a body's cancer cell or as bad genetic engineering of the human body. Murderers, rapists, exploiters, warmongers, and sociopaths are classic examples of improper natural conditions, something that I call the "ideological immune deficiency syndrome." In the physical body structure when the molecules are not arranged in an organized structure, it cannot prevent infection because they are not complete to form a defence. This means that the blueprint or the genetic engineering of cells is not in line with proper natural ways. On the religious side, the nonbelievers are the stray molecules (people) that prevent the spiritual system from working accordingly. In the physical body structure when certain molecules fail to sound the alarm when bacterias are invading, we can become very ill and can die of infection. When certain people with misunderstandings prevent a good ideology and religion from taking effect in its created course for human behaviours it produces the deficiency syndrome and creates murderers, gangsters, rapists, exploiters, and heartless people. Nothing else but an ideological religious immune deficiency syndrome.

As I said, when an individual has too many problems and ailments it means that many of the human body cells are out of condition, mostly because the human mind lacks the proper understanding to prevent it. Molecules by not having a proper connection or being charged with proper chemistry and thought cannot sound the alarm when bacterias invade because they are not equipped to do so. In our ideological process many governments and people also prevent proper ways from blossoming because they have gone astray into superiority complexes, exploitation, and hatred and are ambitious to the limit. Failure, then, to be charged with the proper spirit, motivation, and ideas are what cause that ideological spiritual immune deficiency syndrome. To be conditioned by wrong methods at an ideological level can create, for one example, weakness of the human body. A weak person is lazy in his improper nature, and this ailment prevents him from having consideration for others and he can very well accommodate himself towards accepting that others will have to work for his personal sustenance.

Weak individuals certainly do not like any ideology or religion that addresses them in proper ways, because they are not strong enough or equipped to undertake their proper duties. This situation prevents these people from resisting an ideological standard towards a reasonable way. Ideas of world conquests, high ambitions, and ideological immaturity are also what can very well prevent that good nature from coming up accordingly. Logically, if a person has "stray" behaviour, such as a superiority complex, or has goals to conquer nations and the world or has high ambitions, he will never or rarely accept a moderate ideology and a way to "share," as all good religions in the world tell us to do. Many people with a "benign" behaviour see the good side of this world so they strive to obtain an ideology for good conditions, but once they fail because they are a minority, another time they are overrun by dictatorships and misunderstandings. This prevents their promotion of logic and reason as just causes to be our human ideological and religious leading star. Our Mother Nature is only concerned that everything that motivates us is (bad, good, or evil) considered a natural success. It could not have been considered a natural success if our human mind failed to acquire any form of intelligence, understanding, love, progress, or alertness.

So, as we have progressed from prehistory up to today, it tells us clearly that the pains and sadness involved in history were a way of nature. Obviously, if nature offers chances to move in some direction, then to keep things, as history and prehistory tells, is not a way of full advancement and/or a progressive manner. It does not necessarily mean, then, that because in the past, according to history, our ancestors have been involved in conquests of other people we still have to apply the same methods. It is not so smart either to keep most of the religion of the world in a state of prehistoric thought. Our world means advancement, blossoming, recycle, maturation, rotation. When people prevent an ideology from either recycling or advancing to come forward they stop the "growth factor" or advancement, creating confusions and ailments. It does not necessarily mean that because certain people feel to carry on in their ideas all the others must follow, because what is attractive to them can be totally the opposite to another category of people. Our human body is a cooperative effort of cells with different tasks. Our human world is a cooperative effort of people with different tasks. In our human body system cells work in a very "civilised" manner because they (the cells) do not steal from one another, will not interfere with cells that carry different tasks, have no ambitions to form either, for example, a big leg or a smaller one or form only the heart, but they all work in a very functional design and synchronicity. In our human world . . . well, things do not seem to be in such order at both mental and physical levels. We are astray at ideological and spiritual levels. We interfere with proper religions, proper ideologies. We steal from one another; we kill and destroy; we prevent the many ideologies from coming forward because we do not understand the feelings and inclinations of other people and things they love. Many prosper on somebody else's shoulders. Gangsterism, extortion, gambling and intimidation, unfair gains and exploitation—it seems, then, that our human mind before it can behave accordingly still has a long way to go, doesn't it?

Since that first amoeba in our world, millions of years ago, everything has been a gradual growing process. The many amoebas, that first form of life, have grown today to be trees, animals, and somehow humans, too. The growing process since the beginning of life on our planet has been a union of many cells or organisms in order to form a structure for better protection, motion, defence, and vision (intelligence). Certainly if our human body was initiated by only one kind of cell, it could not have gone very far. Many kinds of cells means carrying out many different tasks, information, specifics, etc., and from here to cooperation in a genetic engineering for an "entire huge structure" such as the body of animals, plants, and us. Information from these tiny bits that are cells enables us to understand our world with its many particulars and to understand ourselves and the world, have visions and wonders about the planets and the stars.

Obviously the many elements originated different forms of life in the beginning on our planet. Logically, as one form of cell advanced, so the others advanced. A primitive carnivorous microscopic organism developed to be a huge unit, such as a lion, a shark, or a tiger. If lions, sharks, and tigers were carnivorous as microscopic organisms billions of years ago, what was the use of them developing so large if they are still carnivorous, as they were in the first place? After all, there is not much difference between a primitive carnivorous cell eating another cell or, in the "modern" era, a lion eating a whole zebra. The truth is, then, that idleness in our world is not what keeps things alive. For the good and for the worst, all living things have to advance, move, and recycle. Our usual business in our everyday life, such as work, education, a career, commerce, having a family, ideologies, religion, etc., is what keeps our human world growing, busy, interesting, spiritually and emotionally alive, an opportunity for passing time, and this prevents idleness. Idleness in our human world can be like frozen water, which can stop our natural growing process at any level (physical, mental, or spiritual). There is not much chance for idleness to develop in our world, because we need food, clothing, and money. We have to grow families and have many problems. Obviously, because our growing process from cave dwellers to our present civilisation has been slow, then our human intelligence became slow, too. Not many really understood that being busy is mostly a pastime to keep a physically aware condition. Precariousness in intelligence, misunderstandings, etc., lead people to kill, steal, and destroy physically, but ideologically and spiritually, too. To deter or criticise a good religion is a way to steal not much different from a burglar entering and ransacking a house. Preventing people from pursuing an ideology suitable to their inclination, love, and possibilities is also another way to steal, kill, and destroy. A leader of state or a superior can kill and destroy if he sides with only that portion of people who share his ideas and criticises or is against other forms of thought and good ideologies. A supervisor steals if he only acts on the interests of the owner, not taking into consideration the status (ideas, emotions, feelings, health, and possibilities) of the various workers. A worker can steal, kill, and destroy (ideologically) if for reward and other personal reasons he accuses and criticises another unjustly. A nation taking too much advantage of the others is another form of killing and destroying. A race taking too many unjustified advantages above another is also another form of stealing, destroying, etc. It is not justifiable that because some of the people like certain ways and ideologies all the rest must follow them. What they like and love can be a nightmare for others. Because of this factor, our Mother Nature across differences gives us chances for changes and recycles ideologies in order to accommodate various inclinations of people's minds. As Mother Nature is a mechanism, we can compare even our human existence to a motor, for one example, and then the various people's motivations, inclinations, attractions, experiences, type of work, etc., are what keep our human motor going or are the parts that make it go. It is understandable that ideologies and religion can be compared to the electrical and gas systems of that aforementioned motor, for example.

15

Language Morality

Mankind in this world was not born to dominate, but to obey. We are certainly not dominant when we are born. Then, willingly or unwillingly, we are carried through that process or maturing, growing old, and dying. If we cannot dominate that process of growing older and dying it means that we are not "dominant." Then we must move according to the natural limits given in all directions. The language we speak must also move within limits, and from here we can raise or build the process of morality.

Morality in our world plays an important role as the principal sponsor of much of human behaviour. Spoken words, moods, and acts embody feelings in people, and as we bequeath physical structures as well as many of the intuitive factors to our future offspring, then logically we can generate youngsters that will go astray.

Let's again look to Mother Nature for instructions. For example, a bird can build a nest without ever having seen it done before. Birds also migrate as adults without ever having done it before. Fish return to very distant places of birth in order to spawn. In the above examples we can understand that we not only can physically inherit body structures, but as well many of the principles of ideas, moods, and behaviours. Our Mother Nature, in order to raise us up in the way we are on the physical and idea level, took millions of years. Let's suppose that offspring then had to begin, point blank, to build up their physical structures as well to feel, learn, and move, then they had to wait millions of years before the birth of the first movement and a functional idea could make some sense. Ailments, brain damage, etc., can prevent proper functions in the physical part and in the feelings, emotions, and intelligence.

A good reason for us to use this is to accumulate proper ways so they can be transmitted to offspring. As we have emotions, feelings, and intelligence that perform separate tasks, as can be seen in the senses, know-how, etc., in the ideological apparatus it can very often happen that a person can be intelligent but has failures or disturbances in the emotional and feeling system. Generally an offspring mutilated in emotions and feelings is rebellious, has no love or consideration, is confused on how to treat parents and friends, and is heartless and unpredictable. This off-centre or stray behaviour can rise because of the unstable present and past genes from parents, grandparents, and way back who had not much understanding for themselves and the world we have and hatred towards one of their spouses, friends, or themselves. They expected or hoped for things very badly but never came to satisfy their will or expected impossible dreams or suffered too much because of an inferiority complex and other weaknesses and a limited intelligence, etc. This could easily be inherited by future offspring and from here create problems. It is easy at a young age to be deluded in many ways, but most of the time the illusions of young people turn into delusions. Delusion began hate. Hate begins to undermine friends, partners, people, the world. Thoughts, ideas, in our brains came from the natural world outside but remain as part of the brain structure and can become part of that genetic ideological and emotional engineering of ourselves. When we marry and have children, these feelings can enter the emotive systems of the offspring, causing them to behave wrongly, if emotions and thoughts of the parents were not aligned with that natural standard. Feeling and emotion "bases" are transmissible from one generation to the other. Our body cells in order to reproduce also have to inherit the new cell geometry and ideas. Thought and geometry in our world are an accommodation of us and the cells to the world we dwell in.

Let's suppose that this world will change from the usual natural pattern we always had; then the genetic codes of living things will gradually change as well. For example, a rose would not have had to develop spines if herbivorous animals were not eating its branches and leaves. A bird of prey would not develop claws and wings if catching prey was very easy and it was not dangerous to remain on the ground. If porcupines had no enemies, they would not have developed quills. If people had no enemies, they would not have any need to develop guns and atom bombs.

A disturbance can very well create a side effect in living things on our planet. As a matter of fact, a disturbance or changes can mutate the structures of living creatures. For example, if we take the "disturbance" away from a rose plant for (approximately) ten thousand years, and the same plant is treated from one generation to another with the same seeds and conditions, then the thorns will cease to appear. Take all enemies from a porcupine's environment and from one generation to another porcupines will cease to produce quills. Accordingly this will take (without enemy) thousands of years.

With the development of the brains, *worries* began to appear on the scene of living things. For example, a snake's molecular apparatus cannot develop limbs and so the difficulty in holding its prey (plus enemies) created worries in the brain cells. Worries then became one more element in the snake's brain to "interact" there. So that kind of worry (now chemistry) created automatically the "how" in the snake's brain cells' apparatus to produce poison to immobilize the prey so it could stand still and would be easy to swallow. Defences are also a product of nature across that molecular mechanism for the better protection and survival of living things.

Understanding now that a disturbance can create good and bad side effects, we can now well understand why there are many emotionally disturbed children. A child's becoming emotionally disturbed is due to an earlier generation (usually grandparents or parents), within an imperfect society, and personal and parental wrong inclinations and failures to understand our created world properly. By not understanding this world properly, our imagination "wrongly" partly lives in our human brains and becomes our behaviour in an unfamiliar shape, far away from natural standards. Having bad feelings and ideas creates bad emotions, which because they are far away from that natural standard module will find the accommodation, among society, nonexistent, damaged, impossible, and it will be difficult to succeed. On the other hand, we also have children emotionally disturbed because their grandparents and parents were treated unfairly even if they were doing the best and good things. In this case, they put into the inheritance of an offspring intelligent and very sensitive dispositions, having a strong apathy towards bad things. With this other case of emotional disturbances it is the society that is wrong and we as a whole or in part should change to accommodate people's feelings to benefit all. Today we pay too much attention to bad emotionally disturbed kids, but we do not even take into consideration those very good kids who were unjustly spoiled in their good emotions. We must not forget that these two categories of emotional disturbances are what produced, on one hand, Muhammad, Confucius, Buddha, etc., and on the other hand, Attila the Hun, Cleopatra, Napoleon, etc.

Morality in any discussion has to be used and calibrated to produce a chemistry limited in a way that helps nature to succeed in our proper balance. Unfair words towards a pregnant mother from the husband or the society or if the woman is self-motivated into bad thought can indeed enter the emotive structure of the offspring. If a father has a strong hatred towards his wife, by fathering a child, these feelings can partly enter the emotional structure of the offspring, originating in many cases degrees of an emotional disturbance against the mother, although most of the genetic inheritance came from the father's side. Generally speaking, if a man has very bad behaviour it tells us that his genes are already off-course. Genetic chemistry can be from the mother's side, and in this case the offspring will behave according to the mother.

Natural balances in our world are another factor to take into consideration. It is not necessarily true that a couple who have no patience towards anything will have offspring behaving the same way. Mother Nature is always looking for an opening to fit us in, creating interests, employment, love, and perseverance where there is an empty space. As our Mother Nature lives with her changes (our changes, too), idleness is not accepted in our world at any level. This tells us that a teacher, for example, will not produce offspring inclined always to become teachers as well, because in the engineering structures of the cells there are also other vocations, beside teaching, which will grow in strength as they (vocations) have a long rest. As part of the brain's and body's cells are well rested, this enables them to become strong towards other careers, such as law, the priesthood, industry, etc. Someone who has been a teacher for generations will drain out molecules dedicated to the good feelings of teaching and will produce an offspring with little feeling for this same occupation. Therefore, as other fields were not touched in the human body cells' structures, then the offspring will be naturally inspired towards these unspoiled fields (the career they like). A hardworking labourer uses too many muscles and/or energies, but saves in mental thought. Obviously, an offspring can be born weaker in muscles' energies but mentally strong. This would open a career, in many cases, in an intellectual field. Mother Nature does not always give us the chances to know what to expect next. We all came from centuries of generations. This tells us that it is very difficult to predict for sure the behaviour and inclination of the next offspring. But one thing is sure, that when we curse ourselves and other people in the world we uproot that natural central line, which in turn gives us emotional disturbance. Also, there is a chance this is transmissible to future offspring, generating their own emotional problems.

Morality, then, is important in order to keep this human world within acceptable limits and with it the trust in our human success in achieving good feelings and natural balances to the world we have, to ourselves and friends and for the good of the family. When mental thoughts and emotions are spoiled then artificial cures such as medicines will become too difficult. An emotionally disturbed youngster is ideologically off-course from that standard blueprint of genetic engineering and finds everything contrary to his thoughts and way of life. For these emotionally handicapped people, the world looks unfamiliar for their ideas. As they are disturbed and away from the real world they will not find an exit from or acceptance into normal society. They will not know how to have good friends because they do not know how to behave or how to speak properly in order to catch the imagination and sympathy of a good friend. Actually, emotionally disturbed adults and children should have a special education to tell them the causes of the feelings of being too weak to work, why they feel like stealing and destroying. They also can learn how to become normal and to tell bad friends the cause of their problems, the use of a good religion, and how to join the advanced society in the future.

Frankly, at the present time we do not have a good society anywhere in the world that can direct people in a good way. Later on, with better natural understanding, perhaps there will be a chance. From past millennia all the good religions have always alerted us to use love instead of hatred, curses, wars, and unfair language. Mother Nature is always behind us, giving that intuitive feeling to use proper religion and good ideas to follow for the success of good human ways. Nature knows very well that when a bad thought is stabilized in the brain it goes astray or becomes rebellious and with it it is hard to find an accommodation in serenity, because the said serenity can only come by following determined natural, good conditions. A stray idea when it is in the human brain tells us to do stupid things with the conviction that we will catch people's attention, make friendships, and receive a pat on the back. Unfortunately, this seldom works without causing problems, or if it ever works, it can only be with friends who have the same problem, but eventually not much can be learned from too much straying. As long as that natural order is not spoiled too much, this human world, even by having good and bad people, can still succeed in giving us a human world within reason. When we fail to respect people's inclination in ideas on their own plus a good religion as a natural cause for good human behaviour, then bad and good people will find too much contrast among themselves and that good nature is pulled astray from its foundations creating bad side effects in people. The good people, for example, can end up with ulcers, heartburn, nervous problems, heart problems, etc. The bad people will crowd jails, because they will be addicted to drugs, alcohol, money, gangsterism, and exploitation, killing to get their way out.

16

Animals and Plants

Because of the type of world that we have and because we are required to move continually at physical and ideological levels, then animals and plants are one of the reasons to keep us motivated in terms of admiration, fear, and proper care of them. Anything that fills our human minds accumulates *wealth for the spirit.* Without the various animals our world would appear lonely, simple, and with much less meaning to dedicate our time and imagination to. Our emotions, feelings, and ways of becoming interested did not come from nothing but were stabilized by so many factors. Therefore, we still must keep those factors that have contributed to the building of emotions and kept our minds active. As long as new generations will be born in our world we need a proper number of various animals and plants. Loneliness in our world is intolerable. Aged people who live alone can easily tell us that loneliness is not a thing to have or experience. If we one day, for example, will cause the extinction of all kinds of animals and most plants then we lose one part of that company that prevents loneliness. This world then will appear to us as quite frightening, boring, and less attractive to live in. We will feel that way because, as I said, we will lose one of those natural pillars that make us feel alive, in the company and knowledge of other living creatures.

Let's suppose that we are in a forest with many kinds of animals. The vision of a beautiful bird will make us wonder. We will fear certain animals, and our heartbeat will increase, creating an emotion. We will feel strange listening to the various sounds they make. We will feel amazed at how certain insects organize themselves. We will fear certain snakes because of their poison, and we keep our minds cautious. Now let's imagine that we are in that same forest, but with no animals of any kind, and we can easily understand that we will lose that wonder, fear, amazement, curiosity, and alertness. The two examples are like life and death. Imagine visiting a busy city, then visiting a large cemetery. We immediately find the answer to the question of why the creation of animals was necessary on our planet. Day and night all over the world, something other than humans must be alive. Many birds will move and sing during the day; others will be active and sing during the night. The sounds of the crickets will make the night appear alive. The bark of a dog, the sound of music, the noise of motors and people will make us interested and with that we feel the emotion of being alive among other living things.

Nothing in this world is magic. Animals need food and space. If they are an aid to our human world, building in us part of our feelings and emotions, then it is our duty to give them their natural habitat and their natural sort of food. Unfortunately, in our present days of immature human intellect we kill animals indiscriminately because we do believe that they have so little to say in our world. We do not understand where emotion, feelings, and interest come from in a nonmagical life such as the one we have on earth. We trap animals; we kill them for fun and ambition, using their skins. We pollute their water; we destroy their habitats in the land and forests. We fail to recognize that they are one of the pillars that build and sustain our emotions, wonders, fears and food chains. There are animals to be kept for food, others for wonder, others for company, others for fear and admiration, others for evil and mystery. Each one has a place in this world for a very good reason. If we one day will not respect these reasons, then our human feelings will fall low and become nonexistent, and with this the will to live will also lower, leading to apathy towards living in the world we have. Apathy to life in this world can raise frustrations and destructions or lead to spiritual death if people one day will abandon the ways to wars and destructions as well the good proper natural ways.

In our world everything is here for a reason. It is better if we study each living creature and begin to respect its reason for existence. Also, our world is made of balances and nothing has to be kept in excess, nor do we have to give away our assigned space. Neither can we take space all for ourselves. Reason in this world has only one meaning, and that is to keep our life moving ahead. Behaving contrary to those reasons, we go astray like a cancer cell, and from here leads to problems, pains, and death. It seems attractive to eliminate all animals that we do not use, as well as take all the forests away to have more space for houses and humans. But unfortunately, we have a nonmagical world and what really caused us to be what we are was the togetherness of all living creatures. Taking away those reasons that grew us across past millennia will cause our downfall as a human race and from here to lack of reason and a stray malignancy in all human structures, physical, mental, and spiritual. New generations will always be born among the adults. Therefore, we have to keep our natural world in shape and order to properly stabilize their (the young's) feelings, emotions, wonders, love, and fear. This does not necessarily mean that because the adults have emotions, feelings, etc., then they do not need to keep certain things anymore. Because new generations need that same sane conditioning that made the adults become as they are. Our world lives and exists with its differences. We cannot pretend to build a motor with only the carbureter and then expect it to run.

A motor in order to run properly needs many parts. There are many varieties of plants and animals. We humans too must depend on varieties and parts in order to form our human body and/or feelings. This also leads to different inclinations between one person and another. The variety of plants then can build an attraction, love, and wonder for the person who sympathizes with one or many of the said plants, by flowers, shape, or design. Having love, attractions, wonder, and admiration increases wealth at a spiritual level, and we feel alive, fulfilled, and happy to live in our world.

Different plants, then, are to satisfy the various inclinations and fancies of people's minds. If plants only produced one kind of flower, fruit, design, or colour, then this could only satisfy one portion (or none) of the people on this planet, leaving the rest to their boring and empty world.

Again, our world is not magic and they (plants and animals) need space and natural respect in order to exist and to grow properly. Anything that constitutes a goal and an attraction and love forms the will to live and the spirit to exist. The energies to move, the muscle cells, come into being according to conditions. Mother Nature has to keep us feeling a fair amount of emotions as well as keeping us physically always on the move. Misunderstandings lead us to undermine "motivation laws" and to use fertilizers and medicines to grow plants and/or domestic animals quickly, cornering us against that natural growing factor. Our body's emotional structures need religion and a rotation of ideologies accordingly. For the physical mechanism we must continually move or be busy in order to carry on from one generation to the other the "will in strength" (muscle cells). It is not in the best interests of the good nature and ourselves to stop this useful natural condition and expect that everything comes easily or use medicines and much fertilizers in order to obtain easy food and with the least work and expense. Nature has its central line and a well-balanced mechanism. Anything that we deviate from, on purpose or through accident or negligence, will not work out but create a painful situation on physical and emotional levels. We can use a lot of medicine and/or fertilizers and other artifacts to produce easy money and food, which in turn gives us more of an easy profit, but we pay out in order to counterbalance that central natural structure, that we have deviated from, in terms of less energies in body strength and less "will" to happiness, to live in our planet on the emotional side. The uses of any fertilizers, medicines, and other unnatural ways to grow animals and plants must be brought into balance in such a way as not to be too much off-course from our duty to Mother Nature. The only way to maintain muscle cells in animals is to ensure they are continually on the move or on the run, which for us human beings means to work. Attempts to obtain easy food and easy things are a sign of mental immaturity. Always keep in mind that there is also a limit to the amount of work we have to perform, because too much or too little work can be another cause for deviation from that natural central standard line. The thought of finding a way to live without work is as old as the world. We remember across history the pains and tears of slavery, the humiliation of losing wars and working for the conquerors. A continual thought of avoiding work has created an existing psychological effect on the human brains, and very many people today have weak feelings towards working every day (the ailment of laziness). This has created those bad side effects such as the exploitation of prostitution, drug pushers, stealing, killing for profit (gangsterism), pursuing an ideology suitable to their unfair conditions, evil tactics, and stiff competitions to reach high places and easy money and easy work.

It is understandable that Mother Nature has ways to advance us. So far, so good, but we always must keep in mind her limits and her extent. For example, our earth is divided into forests, farmlands, rivers, lakes, roads, and towns. If we do not want to have limits, it would be as if we wanted all the world paved with asphalt for roads, eliminating all the farms, forests, lakes, etc. Also, if we want only "one ideology" for mankind, here too we will pave over or eliminate the many other useful factors that are making us feel, be motivated, and have particular inclinations in life. Many plant varieties must be kept for air-cleaning purposes, beauty, and emotive attractions. Forests also should be kept in that normal amount in order to limit too much of a change in natural conditions. Quick changes in weather conditions can possibly mean changes in the molecular structures and other values in the food chains. Our human body has billions of cells that depend on those food chains, as has been familiar across millennia or millions of years.

Changes in the usual food chains also create changes in the cell structures and awareness, changes not necessarily for the better of humanity. If in any case we want changes, it is better to wait until we have better knowledge of the world we live in. A good soldier during a battle if out of direction from the command will stay in a safe place and will come out only when the situation is clear enough for him to act. In our present eras we are going too far, with little understanding. If we change unnecessarily natural conditions a little bit here and there, then one day this human race will end up becoming addicted to off-course changes leading to causes not really tolerable and pleasurable, then this same human race will start to raise the question if living in this world is worth the effort. In deep despair and frustration, the people or the entire human race will no longer search for solutions but to terminate our existence.

Billions of years ago, at the beginning of life on our planet, cells with different properties and mechanisms gathered in a community in order to have or to make a better and more advanced mechanism as our human body is today. Cells have the responsibility of keeping us physically fit as well as building ideas and advising us when we act against nature and when the existence of the human body cells itself is threatened. We want to ask the question, "How do cells succeed in advising us?" ("How do they do it?"). Well, the motor, energies, and geometry of cells have been accomplished across millennia from generations to generations according to necessity, survival, awareness, and discomforts. Keeping in mind that cells are functioning engineering, because another "greater mind," which created the planets and the stars, arranged a natural world to become so. That name, which will be highly loved and respected in the following centuries, is God Almighty.

At the present time, as we all are part of that natural assembly line, we still lack many parts, and this makes it difficult for us to have a sane physical body and a happy spiritual and well-founded emotional life. When all the problems will be levelled down in the future generations then life will appear smooth, with love and serenity, and therefore that apathy towards God that we have today, because of the many problems and pains, will mostly discontinue and people will think of God in different good ways.

Again, speaking of our body cells, we cannot act foolishly and change their structures across any of the man-made artificiality. For then we set up a world for them that simply does not exist and the feelings and emotions, etc., that they produce will never find satisfaction and a pattern to lie on. In other words, if we follow an idea that we have in mind but it is untouchable and nowhere around, then we will live with a sense of unfulfilment, frustrations, and apathy. It is then logical and natural to follow only possible dreams that find or can turn to reality instead of impossible or nonexistent dreams.

Our human body cells are not in control over our human minds because the brain is formed and engineered by the cells, but the mind is formed by visions, the stars, people, needs, animals, danger, beauty, all part of the process. If cells do not have total control in everything, it is then obvious that our learning process from adolescence has to be equilibrated towards a satisfactory central condition without many impossible dreams and stray behaviour either towards morality or the natural offered world. For example, if the animals in the circus will be tamed to be good they will be good, but teach them the opposite and they can damage and kill easily. If the human race has always been in wars and killings, it frankly means that their way and our present way as well are lacking proper learning and the accommodation process and from here to all kinds of madness is a shortcut.

As I said, with Confucius, Buddha, Muhammad, Moses, etc., our Mother Nature gave us the best instructors, but as human intellect could only grow gradually, then the human race really did not understand much from these prophets and so problems and pains have always continued. Our human body is made from many elements because of the various structures that carry on as physical ideas, feelings, emotions, and soul, and it is easy to understand that with a human brain that develops slowly in intellect then things are difficult to master, to discover and apply. It really takes thousands of years only to gain a little information about our world and surrounding space. History tells us just that. Our natural offered world is not without foundations and goodwill if, over centuries, this same world has always attempted to direct us towards a good human existence with various religions and prophets. So it is not our Mother Nature who is heartless, but it is our rebellion and misunderstandings that make our world too evil and with ailments. Across the centuries there always has been thought to defeat Mother Nature instead of obeying her. In our present days it is not much different. By defeating nature, we defeat ourselves, because we put out of condition the existing mechanism that makes all life move and live. To help and follow nature is an act of high civilisation. To defeat and disobey nature in many cases is possible, but when nature is defeated, things will not work out and frustrations, ailments, and pains can easily arise.

Since the beginning of life on our planet, our Mother Nature always prepared a way or a plan to bring up life as we have it and see it today. Behaving contrary to that process, we can form or rouse in us an emotion or a spirit or a feeling that will not find an easy accommodation in the world we have. A person in a bad spirit (bad behaviour) is astray from that good natural plan and can only kill, steal, suffer, and destroy with murder, rape, and wars. Cheating nature does not work, because we must have a central line (as physical and spiritual). Otherwise we could be easily led astray, and with that comes the end of that good molecular engineering process that forms us. It is understandable that at any time if this centre is moved, things begin to tilt in confusion.

17

Limits and Extent of Mechanization, Computers, Television, Radios, Movies, Etc.

We are very well aware now of that existing natural road or central line. As things are, we can only drive in the assigned space of that highway. If we do not, various accidents can occur that, translated into human terms, mean ailments, frustrations, pains, and death.

As the world is not magical but physical, then we should not expect that one day mechanization can replace all human work, because the principal automotion for human cells' awareness is goals that we must continually move or work towards as the only sort of realization for muscle cells as individuals and to future offspring across our genetic codes. Work, mild competitions, necessity, fear, ideas, etc., make us motivated. Note that the offspring of herbivorous animals walk almost immediately after birth, because of the constant danger of enemies that they face, and thus the extra effort of nature to build in them stronger muscle cells. Instead, human offspring require a year, in many cases longer, in order to take the first steps. Eventually, we humans for the last seven thousand years have learned to fight away the many beasts in the forests, and that caused less necessity to move and run, with this weakening the building of muscle cells in us and future offspring. Necessity creates automotion. Security then causes less fear and need to run. Less movement created weaker muscle cells and from here the longer time for human newborns in order to walk erect. Deep in the past, perhaps twenty thousand years ago, human offspring were able to walk in the time period of one month because of the dangers of the wild and other necessities. Baby dolphins because of the necessity to breathe air swim right after being born. Baby birds, for example, do not have to fly soon, because the parents have the possibility of building a nest in a safe place. If the dangers are greater, then "time" becomes faster, speeding up their achievement in moving and flying. Necessity, fear, etc., can cause a faster way to move and from here the chances to build muscle cells as well as awareness and the understanding of what is good and what has to be feared and avoided. Certain animals and plants have also learned their own accommodation so that, for example, when they are where rain is very rare they grow and reproduce in the matter of a week just after it has rained. As we can understand, then, that danger, necessity, weather conditions, pains, etc., have formed awareness, adaptability, and muscle cells in animals, plants, and ourselves.

Again, from these above examples, we cannot expect that one day mechanization will totally replace our daily work forever. For then people one day will lose that natural process that stabilizes our emotions, awareness, etc., and can become weak and lazy so that even to take a teaspoon from the table can be difficult and unpleasant. On the other hand, computers also cannot take all of our mental work away. Otherwise our minds will fail to exercise towards thought, leading to the destruction of emotion and spirit. Death of the spirit for future generations can motivate that nostalgia for a time when wars, etc., motivated people's emotions and heartbeat. If Mother Nature in her plans included mechanization and computers it is because we have to develop a basic civilisation in order to look good. Completely lacking mechanization our present civilisation could look empty, poor, and not much different from other animals that share our same planet. According to nature, then, mechanization and computers are useful, but only to that limit and extent in the natural, proper process.

We also must keep our morality "green" and fertile with good religion and ideologies. Those connections that play an important part in our human behaviour such as television, radios, and movies will have also to orbit in a way to produce a "green mentality" in our human race. Any word, address, advice, teaching has to be managed in a manner that produces ideas and acts according to that natural basic process. Going too far in the ideological factor in movies, TV, and radios can create a stray mentality that can take the "green" away from our human spirit and serenity and replace it with certain emotions and feelings that are not in the best interests of that good humanity, (rape, murder, hate, stealing, addiction, unfair treatment to others and self, emptiness, frustrations, no peace with oneself, unnecessary fear, creation of certain beliefs not fully existing). Bringing to light certain humiliating human realities and conditions can degrade our intelligence and our pride as living beings in the world we have (horror movies, movies that show merciless killings, rape, misery at any level, the showing of human bones in museums and pictures, and so on). We must always keep in our minds that the beauty of a tree is to be covered with leaves, flowers, and fruit. The beauty of human life is to be covered with whatever makes our world look pure and with pride, as proper religion and morality can do. A tree without leaves, flowers, and fruit will look like a skeleton. A humanity without a good religion and morality is spiritually dead or an "ideological skeleton" or a spirit that in this world will never find proper accommodation and peace.

Our natural world since its beginning, billions of years ago, always gave us the same things, such as water, oxygen, the seasons, the elements, heat and cold, the necessity for motion, fear, frustrations, etc. Keeping these factors across a long period of time, nature succeeded in obtaining all living things, including ourselves. We can change things from that natural pattern, but then we can end up with bad situations for which we have no cures. Eventually, when things are astray from that good natural design, there are not many cures or corrections. If our world were magic, then we could correct any situation, but unfortunately, we are a mechanism and the only way to keep it good is to understand nature and follow it accordingly. For example, Mother Nature always has advised us to use a good religion, since the beginning of man's civilisation. We, with the very limited intelligence of understanding, opt to have human laws that result in killings, jail, wars, and an idealism based on accommodating people's feelings, not always in a good natural structure and good faith. All these factors have created the ways of war and destruction as well as all kinds of ailments. Man, since all these problems began, has never really had any answer for stopping either ailments or human killing and pain. We do not have many cures for everything today either, because the said cures for the ideological or spiritual side are only prevention. Personally, I too wish that all the ideas we have could become true, but this is not possible in the natural world we have, because it is only a mechanism. Many people in this world and nation persist in endowing us with certain ideas and ideologies at all cost, not at all aligned with that natural central growth, and then the whole nation and the world seem only to be lost, in confusion and pain. Too much use of mechanization shows clearly that there is something out of order in our human civilisation, such as too much ambition in a world that can only offer moderate conditions, weakening the human spirit and patience and from there causing the downfall.

It seems that we in this world are playing God and mostly take for granted that this world should obey our will and wishes, when in reality it is the other way round. We forget that God and Mother Nature sponsored life billions of years ago, not us. Therefore, the uses of mechanization to gain easy money, forgetting our cells, the employment of people, our awareness and spirit, a way out to move, etc., is not a civilised way.

Another form of hooliganism against nature is the use of radio, television, and movies for purposes of gaining only money or to obtain unnecessary revolutions and ideas, hate, and accusation towards people and nations without proper reasons. These are also choices not aligned with natural realism and the proper human spiritual design. It takes too little to spoil artificially or prematurely the coming or blossoming emotions of the young for commercial gains for those people who are behind the scenes in television, radio, and movies. Once again we are playing God for purposes of gains, forgetting and destroying proper emotional growth in us and young generations, causing confusion worldwide. In a created world we must learn to obey proper natural laws and not play God's part, expecting that this natural world should obey all our wishes instead of moving and behaving only within the limits and to the extent we have been given. In our world a crippled individual has a limited adaptability in the world we have offered to us. A human mind can become crippled when behaving contrary to the natural offered world. Many body cells can become crippled if the individual is an addict, overworked physically and mentally, lives with much pollution, bad ideas, and other ideological unfairness, uses a limited variety of food or uses too many fertilizers and medicines to have easy food and other things. Badly damaged cells can reproduce physically incorrectly. They can also carry wrong emotions, feelings, ideas, etc., which this bad spirit can lead to murder, rape, homosexuality, prostitution, easy gains in "unfair commerce," etc. Having a mentality inclined to stray goals can create a missed ideological "chromosome" or feelings leading to the cause of or the formation of a depressed spirit. It does not matter if the person is either rich or poor, beautiful or strong.

Eventually, a crippled cell interferes with the normal task of the good one and we feel the discomfort. The more crippled cells we have in the human body, the more discomfort we have. There are many varieties of cells in our body, and as long as they follow proper natural geometry, ailments can be avoided. Medicines can cure most of the physical ailments, but when the problems are ideological and emotional, then medicines will not have much effect. Therefore, only a good ideological and emotional teaching becomes a proper medicine. Confucians, Muslims, Christians, etc., tell us that, but unfortunately, we still do not have the scientific knowledge to believe it and there is much confusion in everything.

If we had a world with more opportunity than this natural world offers, then maybe we could accommodate the many deviations or ailments from the proper ways and growth, but unfortunately, we do not have that kind of world. We could accommodate the many people who want things their own way in the wrong pattern, but here too, unfortunately, our world is not magical and the wrong ways do not have another earth and another nature to fit into.

Again, we could ideologically grow a civilisation inclined to kill just for fun and sadism, but then who would be the guinea pigs? We can grow a civilisation with the only goal to be rich, but then we will cripple the ideas of other people who see other values in life rather than only getting rich. We can grow a civilisation with an ideological inclination to never change, but in this we cripple that natural process of recycling for an advancement and inclinations in many other people. We can have a civilisation with the nonbelieving in God and Heaven, but with this we cripple the feelings of many people and intelligence alone will not be able to cover up the spectre of the grave as an end to life and this alone can be painful and depressing.

So we must learn to forget certain dreams or ideologies when the realisation of them, even if possible, would never work for that good human spirit. We can compare our Mother Nature to an automobile. Both are mechanisms. We can change nature, as we can throw away the wheels of the automobile, but the reality remains that then the automobile will not run. If we help to make a better nature or make better wheels, then things will run better. Nature pays and rewards only those who obey its laws. As the laws are always the same, so nature does not have any means and possibilities to accommodate all human wishes unless by limits of that created plan. If limits do exist, then there are limits also in the uses of mechanization, TV, radios, movies, computers, etc. Too much off-course in human behaviour caused less bonds to natural design. It has to be understood that the various cells in our human body did not really engineer everything by themselves, but other factors helped also in the development of the process. For example, the feet that we have flattened because of the body weight. Weight then is another factor that enabled cells to pursue their body engineering.

18

Destiny of the Nations

If our human body is a cooperative effort of many different cells in order to become or form what we are, it can be also understood that the various nations serve the purposes for an ideological mechanism and for prevention against certain human and national behaviours' necessities or possibilities. If our human body was formed by only one kind of cell, then this put out of existence other cells that form our eyes or the mind and heart, for example. If this world was formed by only one nation and race, then our ideas, learning process, and spirit could have suffered a severe blow. Variety, then, is the forming process of our living design. Again, then we must leave intact any variety that serves the purposes for that enrichment and success for our physical body and mind.

Looking to the world we have, there are many kinds of plants, animals, and races. Everything is to avoid that sense of loneliness and boredom and be away from a world that could have been always the same. Each form of living thing, big or small, creates in us wonders, amazement, a gentle thought, a feeling of love and protection, fear and frustration. All that makes in us that sense of existing, being alive, being busy in something and having that sense to break away from a silent and meaningless world. Across the millions of years mankind dwelled in the forest, bit by bit he accumulated physical feelings of existing and being alive and with them motivations, interests. Today that sense of being alive is the result of many, many events in time and synchronicity with the rest of all living things as well as necessity, fear, weather conditions, frustration, old and new generations, bonds to and love of things we like, old and new generation gaps. The many nations offer many pieces of difference to feed mankind with a variety of wonders, history, types of ideas and civilisation, race and colour, importance, mood of people, survival, weather conditions and plants, languages, occupation. Each difference makes the enlargement and enriches people's minds.

Let's suppose that this world had only one nation, one race, and one kind of animal, could we then have our ideas and ideological and emotional factors developed? Only if we had a magical world! In our world that magical factor can only be pursued by physical reasons. These reasons are then the substitute for magic. Before undertaking any movement, we must think of reasons, then achievement, success, but must also be aware of the side effects to oneself and the others. Reason, then, is the magic factor that keeps us alive and civilised and with a human face. Good reason keeps nations as well as a variety of ideas and ideology, differences in languages, culture, costumes, music, ways of teaching, and the learning process. As our magical factor in our world is reason, it is obvious that our ancestors, when the development of reasons were scarce and uncertain, found their history full of pains and death. Eventually, any way that our ancestors turned was a natural process towards the achievement of more and more reasons. There have been conquests. Nations have conquered other nations. Empires rose to the scene, and also slavery and superiority complexes. Revolutions and unrest followed.

Many of us today find our human history crazy, which

is true, but as human reason can only develop and grow gradually, then it had to be expected that the many wars, madness, and bad treatment of other humans could have been nothing else but unreasonable. If reason in this world could have been produced at once, then even before the primitive cave dwellers man would have had all kinds of universities and schools and all would be reasonable, good, and very peaceful.

For one more example, we know that patients in mental institutions have very limited use of reason and thus all the bad and weird things that they do have to be expected. As I said, then reason in our world is a kind of magic and all future generations will have to depend on it. Perhaps in the future there will not be any more blind ideology and blind concessions and behaviours as well as critics but accept only that good natural process that makes us succeed. For the present we have to keep an open mind to the idea that reason can be held in the human brains, but still it has no magical power. Habits, hate, blind hopes, and laziness, etc., can raise or cancel reason from our minds totally. People who can be bonded or keep reasons strong can always and as human rights choose an ideology separated from anyone who has gone astray. As time passes, human understandings will better develop, so we can begin to put in place the various natural puzzles and understand the rules and laws and purposes.

For another example, if we look to the world, we can easily see that there are large nations while others are so small and we quickly think that there is something wrong on the part of Mother Nature in not demanding equality. Could a small nation defend itself against a large nation? It seems that there is no chance. According to nature, there should always be reason! For example, an insect is eating a plant, which has no defence for itself. Suddenly a bird comes from nowhere and eats the insect; that plant is defended.

Let's go back now to that tiny nation. Can anything come from nowhere to our aid if we are attacked? Finding the reasons, we can solve the natural puzzle and we can clearly see that any nation, small or big, has its group to which it belongs.

Take a fast look at nations of the world. We find out easily that there are continents that have nations with the same language, more or less, as well as religion, colour or race, behaviour, etc. This tells us that nations have their groups where they belong. Also, the geographic world seems to be distributed equally for each race. North America, South America, Asia, and Africa have been the parts of the world where the human race moved from one place to another across the centuries, generating ideologies, ways of life, and evaluation for better understanding, even if these things spread through wars or other painful methods. In fact, as our Mother Nature has a central line, balance, and reason, then it always created structures, equality, accommodation in terms of territory, wealth, languages, and religion. In terms of territory, religion, etc., there is distribution between North and South America, the Muslims, Buddhists, Soviets, and blacks of Africa. Nations then are branches belonging to a "tree" (which has balances, functions, and reason according to colour, race, religion). Therefore, when a nation has any kind of problems today or in the future, it is up to the nation and race where it belongs to come with the necessary aid. As our human emotions, movements, advancement, etc., can only come from a mechanism of differences, it is then obvious that we must have many nations and races, with many languages, costumes, etc. To have differences for motivations is good nature, but this natural buildup needs protection in order to prevent one nation or one race from conquering or destroying with habits, boldness, ideologies, etc., all the rest, and then that good natural mechanism of differences could have suffered a severe blow. Nations and races have attempted, across the centuries, to conquer all but, in reality, never get very far. The reasons are that if we succeed in having all the world as one nation, with common laws, costumes, and languages, then we will have crippled severely the proper formations of human global emotions, goals, distinction to keep clean and civilised, advancement and awareness towards respect and consideration to another human being. Understand, then, that human understanding and consideration, awareness, and a good heart can only be produced or obtained across many parts. It is good to know that if only one ideology and one nation could have been all the world we have, then it could be very possible that if there was a wrong interpretation, wrong emotion, laws, and feelings, etc., then that was it, meaning this had to be kept forever because there was no one and nothing that could give advice or a rebellion to bring us to sanity. One of the reasons certain races and religions do not go far in bad things is the knowledge that on earth there are other races and nations that by instinct do not follow madness. But let's suppose there was one nation and one law. Whom, then, could they fear? Certainly no one, and then this human world could have been lost or escalated into weird situations of all kinds. If we do want to play any game, we need partners. We cannot play, enjoy, laugh, and have emotions, goals, etc., if we play with no one. By having partners, we see and feel other people's emotions and behaviours. We can also stabilize and understand, when we are better, if we are more fair and honest or if we are inferior, bad, and unfair.

In other words, it takes many parts for a motor to run. If a motor could be created only with the valves, for example, could it run? Obviously not. Therefore, the varieties that we have in our world in any directions, plants, animals, races, nations, groups of nations, etc., are parts that made and will make possible a human race capable of having emotions, feelings, love and hate, consideration and understanding, intelligence and ways of advancement, respect, and devotion.

Mother Nature gave us a world part fixed, part mobile, and it is not dictatorial to keep certain things always the same. We should obey accordingly that good natural process that keeps us in a moderate sanity, consideration, an adequate understanding and respect. We know now as a fact that animals, in order to keep their awareness and muscle cell mobility, must have enemies. Logically, if all the animals could keep (supposedly they could understand) their awareness, mobility, and self-exercise for muscle development, then there would be no use for enemies. In our human ways also there is a way to avoid any wars, revolutions, apathy towards the world and fellow humans, laziness and madness, ailments, hate, superiority and inferiority complexes, supervision, commands, etc., if we could, from our own understanding, exercise ourselves mentally, spiritually, and physically according to proper ways. Failure to self-motivate accordingly will produce wars, revolutions, divisions and unifications, rapes and murder, apathy towards religion and ideology as well towards leadership and supervision. Bad ways are and remain the only exit left to keep mankind on the road to awareness, emotions, to be active, etc., when proper ways are absent. So to think that perhaps one day there will not be wars, ailments, hate, and other unfairness without using proper ways is totally out of the question.

All of this can be avoided, of course, if we can bond ourselves in dedication and attachment to a good religion, work properly, and love accordingly, as well as use the rest of our obligations and habits according to that good, natural process for our success. Any failures towards proper ways automatically turn people astray and from here lead to wars, hate, murder, laziness, feelings of superiority, unreasonable feelings and emotions, apathy towards the world we have, rich and poor, poor nations and rich nations, etc. The many differences of nations, religions, and races serve the other purposes so that during a war mankind's minds can be set again towards a central consideration and understanding and from here continue to have a form of good vision, which the proper vision could be completely lost: (1) if we failed to properly self-motivate and (2) if there were no revolutions or wars or at least attractions to bring the proper way back, such as good emotion, feelings, consideration, etc.

If there is no one to complain when things are bad or astray, then our human emotions, feelings, understandings, love, and consideration will fail to form or grow properly and from here many will be lost in frustration and madness while others will just die spiritually. Logically, at any time we disobey proper ways, our civilisation can only go from bad to worse, causing an increase of pain to keep us away from boredom, a null and void world, which could result in more devastating effects than wars and pains. Perhaps it is premature to say that future generations can live without wars and pains because: (1) it is hard to be self-motivated by following proper natural laws, (2) habits are quite capable of confusing reasons/intelligence, making us forget, and (3) when we are in a good civilisation we believe that things will not return to bad ways again, creating a situation of losing contact with facts and from here falling into disgrace.

Still in our contemporary time there are many people that really do not seem to fit a world of differences. Actually, they seem to be the cause of many problems that exist in reality, but without differences we would not have a world at all. This is why nature came with variations of plants and animals, as well as different races, religions, colours, languages, and behaviours. Certainly in our early stages of human understanding it is easy to believe that by eliminating our human differences (religion, languages, ideologies, etc.) we could end up with a more peaceful human world. This could only be possible if we had a magical world. But as we have a mechanism for a world, then this is like pretending that the motor of our automobile should or could run only with the spark plugs.

For another example, in order to produce or grow vegetables, soil alone is not enough if there is no air, water, space, and light. Obviously, if we could eliminate space, water, and light and still could grow plants surely things could be easier, but unfortunately, this does not happen in our world. Now these conditions of light, water, air, and soil are essential in order to have our daily bread, and as Mother Nature always uses the same laws, but with differences, we can also understand that having only one ideology, one language, one religion, one nation, and one race in charge of all would not have been enough to create links that, put together, made possible the way we are, the way we feel alive or to live, emotions, hopes, differences in ideas in our minds, worries, love, consideration, devotion. If we, for example, take away one of our feelings, emotions, love, consideration, etc., then our light or that feeling of living or being alive will also become dimmer and dimmer until that stage of spiritual death. If we take away from the motor the various spark plugs, valves, carbureter, etc., then that motor becomes less and less useful. If we take away from a tree the various elements—air, water, light—then the tree will not survive. To keep, then, nationalities and differences we do not take away any useful linkage from that natural mechanism and we surely will go a long, long way. We must not forget that everything we have on our planet has many purposes. One of the many purposes of nations, besides emotions, awareness, etc., is to play a counterbalance with human intelligence running short of its glory. Human intelligence does not always function perfectly, because habits, laziness, ambitions, and hate can put that said intelligence to its knees. Then the only way left, as in the past history or in any future, to cause wars will remain the only alternative in order to bring back proper ways. Failures, then, to abide by proper natural conditions for human good sense, peace, awareness, etc., will have to cause in the hard way or from the other side of the coin of Mother Nature. As things are, then, we cannot call an "enemy" a nation that tries to conquer back its natural identity. It is not a tragedy if a nation conquers another as long as it is for a short period of time and good reasons. After all, is that what we have learned according to the last five millennia of human history? In a nonmagical world such as the one we have, that "justice" must come from somewhere. Wars are only another pillar of nature that make sense when things are going astray and to prevent this human world really going down the drain. Unfortunately, for many people wars have become a genetic habit, and wars for these people can only serve the purposes of satisfying a sick emotion instead of serving as a corrective way for certain parts of humanity or as a whole. If certain people like wars "only" to satisfy their sick spirit, then they can meddle only with the ones who have the same stray problems. Awareness, intelligence, etc., in the human world have been made possible because of necessity and the presence of danger and/or enemies. Wars also in the future can come for another reason, when humanity will fall to a lowness in awareness, respect, failing to work and grow properly a family, etc. Then as we have a nonmagical world and neither medicines will work in this direction, then fear, laws, and wars will remain the only other alternatives in order to rebuild certain reasons. If wars have to continue, in the future, we must not be abandoned to all forms of cruelty and savagery, but just as painful games become necessary to curb evil and to also bring back the many who spiritually die because of lost faith and love to the world we have. The best of proper ways is a good, natural law and a good religion. Once we fail to abide by it, all our emotional spectrum will shift to unfairness and from here to evil downfalls in all directions, small and big. It is understandable that many people see religion as useless, but the real problem is that in our contemporary times we do not understand how or where our emotions enter the human body cells, and so we are called to believe that material things should make us all happy, and not religion. This concept created a precarious intelligence, and then proper human spirits were weakened by centuries of generations under all forms of misunderstandings. Religions have to be scientifically evaluated and understood so we can apply this very beneficial *natural* law to mankind. We must never forget that our Mother Nature always uses the same laws, but with differences in all living things. As an example, then, if plants need air and sunlight and soil in order to live, then we as humans need ideologies (air), religion (sunlight), and food, etc.

19

Continuing in the understanding of the natural process, we find many "species" of plants or animals that are so similar we can classify as belonging in the same family group. For example, tigers, jaguars, lynx and the domestic cats all belong to the Feline family, just as lemons, oranges, and tangerines belong to the same family. Now by analysing the various nations (nature uses the same laws always, but in differences), we also can find their original core and the species that group them in a specific family. Starting at Central America way down to the tip of South America, all these nations included, there form same-family groupings. North America and Australia form a "mixture" that can be classified as another group. Western Europe seems to be the origin of these two groups. Eastern Europe has grossly contributed to form the other group belonging to another class that takes some part of Eastern Europe and the northern part of Asia. Then we have another race, including: India, Pakistan, Bangladesh, Turkey, Afghanistan, Syria, Iran, Iraq, Jordan, Saudi Arabia, Egypt, Libya, Algeria, Tunisia, Morocco, Spanish Sahara, Mauritania, and other territories. Then we have the yellow and the black races. These six categories or groups that form the human races (mentioned in the Holy Bible as the horses of the Apocalypse) should each group themselves where they belong to set their own internal problems when they (problems) arise, unless they go astray. Then other races can intervene to set natural proper ways.

In the past there have always been invasions of one

group of another for certain reasons. Our human intelligence is a process that can only be formed by evaluating different thoughts, behaviours, theories and counter-theories, good and bad treatment among people, rights, etc., which is what sets the course of civilisation up to our contemporary ways. When things will be better understood we will follow without wars, etc.

Analysing the various group territories seems to be, in a gross way, divided equally. In the case that future generations, either for ambition, hate, or madness, will change natural boundaries, etc., then pain will remain until nature fully recovers.

The facts are and remain that as our planet circles around the sun, the sun appears to make a half-circle across the sky from sunrise to sunset. Then as that half-circle is the principle of geometry, everything is according to a circle and from here to a centre, balances with equality in divisions. We also must not forget that our earth is a globe or a circle generating differences in climate and also different plants, animals, and human races. If our human world was without geometry, then perhaps things could have come in different ways, but obviously nothing can grow without laws. If no laws were in existence and some form of life was developed, then this form of life did not have to obey any laws in order to live. As I said previously, only one race and one colour could have been lost through limits and failures to develop a functional intellect. Nature has its plans. It is not necessary to be involved in wars, revolutions, and unfair acts in order to evaluate a good way of life if we voluntarily accustom ourselves to proper exercise towards reasonings. But if we fail, then that natural mechanism is and remains always there and problems will certainly arise because an ungeometric way is always suicidal.

Our created world is also well balanced, and there are not many chances that today, or at any time in the future, one race will ever take control completely of another or all. Our world has not been created for slavery or superiority of any races or any evil ways, but only to keep a mechanism going in order to temper our emotions, thoughts, feelings, abiding with our fair ways, wonders, possibilities of one race showing good civilisation and ability to other races by pursuing devotions to proper natural and good conditions. Eventually, if there was only one race, to whom could we show our good work and dedications? If a nation shows the world the good things others will follow, perhaps for that creation of competitions, good behaviour that really makes this world turn. Remembering always that competitions must be balanced to human possibility and never must cross that central good line. Stiff competitions can be the root of many evils and ailments. As we now can analyse natural puzzles, we can easily understand that this world has really been created more or less to play games and not to be involved in acts of cruelty. This also can let us all look at the future with more optimism that in case of wars people will not exceed in any form of madness but only for a "friendly lesson" in order to return things to normality. Atom bombs and all other sophisticated weapons can then be abolished. Any armaments also should be equally distributed among the groups and races. Never give to any group or race any form of high profile or superiority or they will think that they are gods on earth. Also, never give races an opportunity to be denied more than the others; otherwise they will be tempted to conquer, harass, and destroy others.

Our natural world since its beginning always has been on the move in its growth process. We know that life developed from microscopic unicellular plants up to today, with trees tens of metres high and huge animals. Races, nations, and families of nations certainly will not stay idle but will also continually move in any direction. As we have a naturally slow-moving world (the millions of years it took to develop life as we see it today show just that), then our human intelligence grew, slowly confusing the already precarious nations by not understanding true situations led past and present civilisation to make our Mother Nature appear as evil and shameful, and also shame is our human history.

It could be useful to remind ourselves that things in this world have time to grow. To move more than what Mother Nature has stabilized can cause behaviour gone astray that can lead to problems physically and spiritually. To have all races mixed is not so good, because by losing that natural identity, it creates difficulty with the proper development of our good human balance, good emotions, intellect, and spirit. Neither is it opportune that many nations will leave the branch to which they belong to unify with other races. Otherwise this can lead to a weakened spirit of the race, as well as to a smaller territory, etc., leaving the probability that they will become subjects or be conquered by other races. Logically, we certainly cannot tie this human world and pretend that things will stay in one direction forever. Therefore, when people in a nation leave their group or race this void can be filled by another nation belonging to another group or race so that, in this way, we will use proper nature in its mechanism, keeping the same territorial balance, leaving also the spirit of all human beings under proper natural determination without being a subject. In our world everything is related. Beginning from the minerals, then plants, animals, bacteria, they form groups closely related to each other. Human races on the various continents are by that natural factor related.

North Americans can be now related to South Americans to Western Europeans and the black world. North Africans, south Asians, the yellow race, Soviets, and Eastern Europeans seem to be related. As we have moved physically and spiritually then our Mother Nature has offered conditions and balances to actualize the said motions. It is up to us to move this natural motor when human emotion and spirit will go astray, become evil, low, and bored. Our entire natural world that we have is here only to give us a way and chance to survive, and it is not necessary to be involved in any extreme acts such as prolonged hate, wars, attrition, and commerce only to rise above other nations or for superiority complex, conquests, etc. Any move that we do make has to be related to natural laws and not according to certain human personal feelings astray from that good central, natural line.

It is understandable, then, that our human nature has not been created for wars, but as human understanding and consideration and devotion can fail because of our nonmagical nature, then it is obvious that war can be useful. If we would not fail to abide by proper ways, things would always be smooth. The problems arise when we fail. On the other hand, our Mother Nature has her created plans and certain ways will have the only alternative to bring proper emotion and understanding back. In other words, then, if we fail to behave properly, things will become painful, forcing our human world to return back to lesser pains and to proper civilisation. This is why perhaps wars will continue to exist in nature in order to mend our good ways back at any time we are directed astray.

People, nations, and races can have differences in costumes, languages, and ideologies as long as they will remain with the race where they belong and will not stray too far from that natural central line. As our human intelligence is a mechanism composed of many little pieces such as knowledge, learned things, experiences, thoughts in many directions, goals, desires, ambitions, feelings, memory, etc., then those parts that condition our human existence have to remain more or less unchangeable. Otherwise, we weaken natural balance structures and from here human intellectual conditions will be put astray towards inferiority, humiliations, and evil. It is not civilised to give too much power to certain individuals, nations, or groups of nations. For then there will always remain that risk of undermining, feeling too superior, etc., creating the illusions that they are a better race or person and from here leading others to unbalances, ailments, revolutions, and wars. We also have to analyse the word *good* and use it to describe human behaviour in its standard natural way. In other words, we have a world that has its "spines," because we must always work, feed, and sustain our families. Cold and heat always will be present. Also, we get old and die. Softening our human feelings in comparison to nature (either physically or spiritually) can lead to frustrations and heartbreak and leave the impression that we live in a world that is evil, difficult, and unfair (lacking adaptability in our world).

In our present days, we learn from one generation to the other and from schools only material goals, and mostly we forget the ideological side. Certainly, money and food are the most useful things, but if our human spirit is ill and bored, then money and food alone are certainly not the answer. Many people have in their lives all they want, but still many of them are unhappy and commit suicide. This tells us clearly that food and money are simply not enough. Religion, ideologies, etc., must always be present in our life with their good natural rules. Looking deep into space we can notice the presence of stars, but there is also light and the immense space. Matter would not have succeeded without space; light and matter would not have succeeded without space. Materialism will not succeed in our human world if not provided with ideologies and religion, remembering always to use proper doses in everything we do. Our human mind is like our lungs; it has to continually move (think). But our human mind can never do good things without purity of the spirit. Purity of the spirit can be accomplished by abiding to these offered good natural laws at our advantage.

20

Looking Here and There to the "Spirit" (Soul), Emotions, and Ideologies

Because in our world we must move in some direction always (good or bad), one of the most fascinating chapters of study of the future generations will be if there is really another life after we die and if we have another spirit (besides our emotive one), which, by excluding our body substances, will transfer, in an unseen existence, somewhere else. From now on I would like to write the word Spirit (Soul) with a large "S" in order not to be confused with the human feelings that include our earthly spirit.

Eventually, as nature is a mechanism composed of many parts, then these created divisions in human theory or opinions exist in order to create interests, attachments, and pride, with this making things with interests, and from here to study and research until something is discovered that is real or theoretically convincing to most.

A good number of people in this world because of our still immature concept of this world will not even consider the existence of our Spirit, nor the existence of another world after we die. Another group of people a little more intelligent will look a little deeper and will at least accept the idea that perhaps the believing of the Spirit for another world is only wishful thinking in order to leave hopes and refrain from that despair in the belief that when we die everything is finished. This second group is, without doubt, much better and well founded than the first group of people.

Finally we have the very advanced and highly intelligent people who feel or sense from the bottom of their hearts that this idea of life alone on earth without another world is senseless and without reality and foundations. This third group of people will be the one where our Mother Nature is succeeding in proper maturation towards true intelligence and understanding.

Obviously, certain facts, theories, etc., can confuse the human mind. For example, if everything is eternal, then there is no beginning or end, leaving no principle for a Creator to create things out of it. If the universe we see has been in existence forever, then a Creator cannot fit in. But this concept cannot be applied in the world we have, because it would not have worked out as we see it without a Creator behind it. It is logical and true to believe that matter and the immense space we see have existed forever, but the "form" of life that we have and our planetary system could not have existed without a Creator.

Let's make an example. We go into a forest and cut down a tree. With it we make boards and build a kitchen table. Surely that table could never have been there if we did not use our skills to make it that way. Logically, the matter that forms the tree or that table has existed forever, but certainly not the table.

Switching now from this example, let's analyse the formation of our planet earth and possibly the stars. We find much truth related here.

As I said, matter and space have existed forever, but not the stars that we see and our earth and our life. As a matter of fact, our human history only emerged few millennia ago but was not there forever.

What else makes us sure that we have been created?

There are many facts that we can analyse to tell the true story. First of all, we are in this immense space that we see because it (the space) gives substances the chances to move in a planetary orbit and rotation and from here the seasons, which give plants the possibility to grow and have seeds for the maintenance of all animal life and humans. Nothing else then but a skillful thought by "someone" (a Creator). Logically, in eternity nothing really can be created, but we can build machines, robots, and computers by using the substances that have always existed. God created us by using things of eternity. Again, analysing the space that surrounds our planet and the stars, we find out that cold is always present. Certainly this "could" have sparked the ideas of "someone" with immense possibilities (God) to use it for later purposes, as could be the separation of small masses from larger one (moons, planets) by using intense heat to counterbalance the existing cold.

The presence of cold and heat could be used for more elaborate laws or to form other things as well. For example, if we have an incandescent ball of molten iron at many thousands of degrees Celsius and we add water to it (water colder than the molten iron), then it creates a strong reaction. We can understand now that the difference between heat and cold helped stars to separate from a primordial mass. Returning now to the principle that when the stars were not there all was only a huge ball with a temperature of multimillions' degrees Celsius at its core but less heat on the surface. This difference in heat, compression, etc., between the surface and the center could have initiated a first, the movement of the masses with increasing motion, and then separation into smaller and smaller pieces such as stars and planets and moons. Obviously, the difference between the centre of that primary ball in molten or very hot substance and the surface there had to be enormous in order to cause separations of parts of the mass. Eventually, as the masses became smaller and smaller because of separation, other laws appeared in existence and the changes gradually of substances into different elements. From here came the beginning of life. Eventually the larger mass remained in their primordial laws in suns or stars. In minor ways, laws that appeared in the beginning of our creation back in time seem now to be applied over and over in our daily life and from the beginning of life on earth.

Let's suppose that we know some of the answers to the question of the formation of the universe, but is this all? It is not. Obviously, if everything has always existed, where were the substances that we have touched and seen before the formation of this universe? In what form? All those millions of degrees of heat did really destroy all those other *laws* that substances had before space and stars and planets? All good religions teach us of another life after we die. Are we then entitled to a Soul?

What we can understand now is the fact that substances have formed life because "Someone" with immense possibility foresaw a Creation in space and also a return in Soul. A mechanic can build a motor, but the parts one day, either by corrosion or friction or rust, etc., will return back to being elements of the soil where they were at first. If things have that immortal behaviour to return always back to the origin, then this fact, this current or force of eternal motion, can be useful to carry or transfer things from one place to another. This concept can build a theory that holds the possibility of transferring a Soul from a mortal being to immortality as a real possibility because "Someone" provided for it. Our world, our existence, does not start or end here. This can explain why we have believers in God and nonbelievers, because our Mother Nature is a mixture of a Creation and eternity. The point remains that the believers are right when they say that this world has indeed been created. The nonbelievers are also right if they go much deeper, to say the space and stars we see show there was no Creator or builder because things have always existed forever. Eventually we live in this part of the world, and as Mother Nature tells us, we must believe in a God and from there a good religion and our Spirit (Soul). Across conditions, in the beginning, God foresaw matter to become earth and from there to life and humans. God also foresaw the ways to preserve in Spirit our life after we die according to conditions here to follow. We know for a fact that there are differences in this world, as can be the immense space that we see and substances. From here we can understand that eternity is not composed of only one thing. We can also understand that all the parts of our eternal world do not exist in the space we see as our universe alone, but also outside it in an endless manner. The words *idea, design, matter, space, dignity, Spirit, afterlife, life, good and bad, light*, etc., already exist and this is why we can come in our world across God's created matter and conditions. In other words, life as we have it is a chance and possibility that existed forever. Substances are a method to let appear the various spirits, dignity, soil, etc., in different forms, one of which is life in substances as we have here. Substances are the mobile parts (wheels); space is the fixed one (chassis).

Obviously, life or to live or the ideas to form life did not have their beginnings in this solar system but were already in existence before any stars and galaxies appeared in the firmament. This tells us that anything that we can think of has always existed. Surely if life did not exist forever there was absolutely no chance that our Creator God could have succeeded in making all living things in our world. Substances, light, air, etc., are visible and tangible, but once we begin to mention ideas, design, space, and Spirit we begin to enter into the ghost world where the only way to see them is to incarnate them or "suit" them with substances and then they become visible and tangible. For example, we design a building, but the building will never be there until it materializes with substances, such as bricks, foundations, rooms, furniture, etc. God had the idea to create our world. God is a Spirit; the idea is invisible, and so is God. If God did not materialize our world, the ideas could never be seen in the way we see this life in substances. Today, on a clear night, we see millions of stars. On earth we hear the singing of birds, the roaring of motors, the winds and the rain, the songs of people. God, an invisible Entity, begins to appear in its presence, and the idea is materialized in stars, planets, and the innumerable varieties of life and living creatures. Our planet earth was thousands of degrees hot before any form of life began to appear. Our earth then had a beginning and will have an end. The sun above was much, much hotter millions of years ago and in the distant future will lose its heat and will grow colder and colder. If our sun continually loses heat it simply means that it had a beginning and has been created, like the rest of everything we see. In this world things that lose their status are things that have been created. The immense space that we see is not losing or changing; there are no clues that it ever will. This tells us that space had not a builder or a Creator behind it, but substances indeed had a Creator at the beginning. Now from what we can see in this world substances, ideas, space, etc., are parts to form a specific structure where everything has its place, destiny, and usable identity. Eventually those our Creator God knew how to resuscitate substances to a form of existence or could let things go in the unknown so no life as we have as well, no solar system and the stars. There have to be in existence other forms of beings that are not in substances and with greater possibilities than us and can be immortal as well. Also, geometry does not apply to their world.

Those Spiritual beings or ghosts do not need a space such as ours in order to live, because they do not need a planetary system, orbits, air, waters, etc., in order to develop and be kept alive. Obviously, then, it is unintelligent to imagine that everything is space and stars, because the said space is only good for us in order to give the chance for planets to move and rotate in order to obtain the various seasons and with this the growth of all plants. Spirits do not need a planet to have seasons and plants, because they do not eat food, and therefore do not necessarily have to live in our world but can exist outside our universe. Psychologically, we see no end to space, and we tend to believe that it is infinite and so everything continues forever only in space and stars. We forget to consider that eternity has no end, no beginning, so not everything stays in one place alone. It is then unintelligent to believe that space and stars are all and everything. If we see no limits to our immense space, it is for other reasons. One of the reasons is that if we could see the end or strange things, then our human intellect, feelings, and emotions, as they are a mechanism, could suffer in its growth because we could be carried away by those strange visions that could develop the minds of humanity to go astray from our terrestrial duties. So we are never able to see the end of our universe, no matter in what star or galaxy we would be situated. Only this way do we feel in our life or in our mind the feeling of infinity, also having "roots" or attachment to earth. Even if we have limits and mortality we have infinity.

Another reason the boundaries of space are out of

reach is because if some other thing could be close to stars, that could put out of condition the orbit of planets on their own also destructing any form of life.

21

Substances do not seem to have great strength and if something else rather than matter were present, then I doubt that substances could have gathered together to form stars and living things that we see but could have been attracted by that other presence and form or be part of something else. Space has a very neutral identity in comparison to the strength of matter. Therefore, it does not matter how weak the power and energy of substances are. They have been touched minimally by space so they can form whatever they can. In fact, our earth rotates around the sun at an astonishing velocity, but space is so neutral that it does not interfere at all with air, clouds, etc., that our planet carries around the sun. Imagine now if space could have some form of resistance, then either air, clouds, or waters could be wiped out easily. Here we can also understand why our Creator chose this immense space to be the "home for matter" so it could reach the possibilities of its identity. Again, with this example we do not see a Spirit (God), but we can very well indeed note God's presence and the incarnation of a preuniversal idea in using substance. The immense space that we see is unchangeable and cannot be modelled, cannot bend or materialize, cannot be moved or pushed aside. Space is not then the Spirit of God but something similar. As space is unchangeable, immobile, etc., it is a good "womb" for matter to move and live in. Also, it gives the opportunity for planets to move and orbit; the opportunity for us to send a spaceship to the moon; the opportunity for that first primordial ball of molten elements or matter to separate into smaller masses to become galaxies and stars; the opportunity for human beings to move in any direction as physically, mentally, Spirit, etc., dust particles can gather in distant places and recycle or begin again stars and planets and so life. Now accordingly right from the beginning of civilisation on earth our Mother Nature always has inspired people to use religion and the belief in an afterlife. Mother Nature is a mechanism, and so too are our intelligence and awareness, which, obviously, come from nature. As nature is a mechanism, it certainly cannot lie or twist things, and it is easy to understand that primitive inspiration from our ancestors, belief in life after death, was not a lie either. Ancient Egyptians left food for the dead to use on the long journey to the other world. This was a primitive thought, as the dead do not eat food, but is an example close to the truth. Our scientific knowledge of today leads us to understand that the immense space that we have cannot be compressed, used, pushed away, or modelled, but indeed can very well offer chances to travel in it. Obviously, then, if our builder or Creator God foresaw the chances for substances to become life, this God also knew that a Soul could travel out of space and in it. A Spirit is an existing creature and needs something to use for its glory and existence style. If a ghost or Spirit could forever be in this space, which cannot be modelled or used for any purposes but substances, then a Spirit could have been either in chains or unformed. As space offers accommodation to substances, it also offers chances for the Spirit to move out from this universe in space and after this long journey find other worlds suitable for uses and glory of its kind (Heaven?). Ancient Egyptians came close to the truth when they left food for the dead so they could use it on their long journey to the underworld. This does not necessarily mean that the dead eat food, but symbolically, the Egyptians had a very intelligent approach to the subject. However, the mystery of how ancient Egyptians could come so close to the truth so early in history could be explained by changes in genetic engineering of cells caused probably by malaria from insects of the Nile River, malnutrition, slavery, the threats of snakes and crocodiles, heat from the sun, droughts, dysentery. All those factors contributed to that natural plan that dwelled in the molecular system of each individual to find a solution to their problems. A solution that was not for their time of course but only for when civilisation is scientifically very well advanced as in our contemporary times. Certainly other tribal civilisations that had those same ailments as the ancient Egyptians could also build natural reasons, telling and behaving according to the same true story and good interpretations. We can mention, for example, the ancient behaviour of the Incas and the Mayas of the Americas.

Since the time of cave dwellers, man was aware of the presence of good and evil. As we have an immortal world where nothing really can be created or destroyed, we must draw the conclusion that good and evil are absolutely not inventions only in this world but have always existed. Our world is good, so then either nature or God could not have assembled a world like the one we have, based on lies, only to let us live in the illusion of belief in another world, just to make us happy so we could live our earthly life with less frustration. To believe that "someone" could have cheated or twisted things around at will, then, we put the existence of eternity in jeopardy and in reality we are reaching nowhere and with no conclusions. The real fact remains that eternity has no limits or boundaries and there is no one, then, big enough to manipulate it. We must conclude, then, that either our existence or the existence of a Creator obeys the same eternal laws. In other words, when we get old there is not much that either us or God can do in order to make us young again or live as immortal beings. But according to what we can see and understand in this world that we live in, eternity is composed of chances and opportunity. From chances and opportunity we can also very well conclude that the possibility of another world after we die is not an illusion but a well-founded reality. Also there is the chance, then, that life does exist on other planets, on the stars, as well as in Spiritual form outside of our universe.

Since prehistoric times mankind has always pretended that the world should obey his will and thought of the mind, mostly forgetting that it should be the other way round. Man has always used religion for divine help, mostly wishing for material things such as riches, fame, superiority, and healings, failing to consider that our existence only offers chances for possibilities and nonmagical and material divine interventions. Politically we use psychological factors to work at a personal and national level to twist things in the ways we want, forgetting that natural central balance and universal laws come first. The "unnatural" results in delusions and failures. God created our world but took billions of years in order to develop life in the ways we see today. This tells us that even a God cannot obtain things in a short while or at the will of the mind. This process also must apply to us, and it is time that we begin to follow those good natural laws that exist in our world instead of personal ideas and wishes that in reality lead nowhere. As we have now the awareness to live in an endless world, but our earth life has limits, then we must begin to abandon the ideas, one by one, of certain personal goals and wishes out of natural reason as well out of reach save for possibilities and chances. It is then civilised and advanced to go with the "winds" of our true existence and begin to accept the ideas of a Creator, the existence of an afterlife, and the believing and following of a good religion. If this immense space that we see teaches us that planets can have orbits, we can carry our physical body to travel in other planets of our solar system and can also send nonphysical things (messages) across long distances. These possibilities have been foreseen by a God not only at a terrestrial and universal level, but in other ways out of earth and universe in that place that since the beginning of civilisation on earth has been thought as a very beautiful place (Heaven). In other words, if everything is immortal, then we have to join the ideas that things are able to move on earth and the universe and also out of this world. Immortal things have the passport to travel anywhere and forever. Therefore, there can be the reality of a Soul that can travel in and out our universe to be welcomed in a place we call Heaven (a Spiritual world).

Now, returning down to earth, our Mother Nature's plan has to continue as it is in order to keep our human body and mind in a functional, stabilized manner. To change unintelligently can cause failures on the part of nature and God to keep ourselves properly not only in our terrestrial life but also beyond. Terrestrial health is an actualization of a well-founded physical condition that should produce enough energies and ideas to be spent to support our existence and goodwill. As for our Soul or Spirit, once it exists it cannot be held in this material world that we have, because as already explained, no one can put in chains or manipulate immortal things so they are "self-motivated" to reach either a very good or a bad eternal resting or "home" place. As not everything in our world is substance, then our human mind is one of the many things that began to depart from that said physical world and so needs infinity to keep its life, interests, and identity. Our human race would have had a short existence on our planet if all the discoveries were only limited to the understanding of the atoms, cells, and living things in general, concerning only substances.

Therefore, as we live in an eternal world where there are no limits, then our ideological growth can go on forever in its thought and imagination, and that will spark in us attachments and interests, a goal to keep on and live, a belief in an afterlife and God, and with this we can build mental energies and serenity, which keep the individual comfortable in the world we live in. As I said, if our mental growth were only limited to the understanding of earthly things, then as soon as this was accomplished our human race could begin to lose interest in life and from then on form no attachment to anything. If we had no attachment to anything our muscle cells would become weaker and weaker, as would the will of the mind. Then as our feelings, emotions, etc., are transmissible, this can also enter the genes of future offspring, making them weaker and weaker, forcing them back in history because of less effort to work, accept responsibilities, love, etc. A good civilisation needs a lot of individual effort. It becomes unintelligent and absurd to expect that God or some other advanced civilisation from the stars should come on earth to help us to accomplish material goals or to explain everything to us. It is understandable, then, that our emotive system and spirit will progress much better through personal experiences, testing, performances, measures, pains of the efforts, being comforted, reassured, and balanced. If someone from the stars comes to work for us and explain everything, that could be good, of course, but an explanation will not carry the necessary pains, comfort, fatigues, awareness, etc., and then we could possibly not learn much out of it. Therefore, it is very unlikely that someone will come to work for us all or explain everything. If "someone" will have to come it will be when we know most of it on our own.

It is not necessary that we discover things every day in order to keep our minds alive and emotive if we can apply a form of recycling of sectors of the knowledge that we have. For example, we can abolish any geographical knowledge in schools and adults for a number of years and then teach the knowledge again. We can do the same with history. We can remodernize ideologies and religion. We can take and add new beliefs; we can use limited variations of music and then open ourselves to variations again; we can limit machinery and computers and then use complete applications again. All this, of course, can keep people always interested. They will have minds with hope and see a light at the end of the tunnel. They will have an inspiration, a goal, and the will to live on. When we possess most of the knowledge regarding the human race, then everything will be recycled till the end of history.

To accommodate a perpetual spirit, emotions, etc., in a world that has limits and mortality we can only use a recycling system. For example, as we die of old age, newborn babies will come to replace us old people. On the ideological side, it is also opportune to leave things open for that changing desire of people or the recycling across many ideologies for that process. The beginning of ideology is similar to a seed. It is born, develops and produces fruit, and then dies out. It is then understandable that no ideology will stand forever but will roll and pass away. We cannot blame people because feelings change, as Mother Nature is behind everything. We try to exclude religion, because we believe it is man-made instead of knowing that religion is a natural process to grow properly our emotions, attachment to life, respect, and goodwill. If we have a nonmagical existence, what would happen then when we cut the supply line for the development of our good Spirit (religion)? Nature gives us a variety of ideologies to better fit our spirit. We also must learn how to recycle them for the future instead of fighting among ourselves in order to have only one global ideology and eliminate the others.

Eventually, if we fail to understand and gather information from the offered world of nature, then history will repeat itself in the old-fashioned way. This in the future can be more catastrophic because of an overpopulated world, and so natural resources, soil, etc., will be exploited, contaminated, and made poor. This eventually will lead to more problems related to the mind and the body. Also, religion is a natural event and it is unintelligent to believe that religion is an invention of the people rather than a natural gift for our own good. It is not good to use a religion only as an idea without attachment, to practise devotion and put a limit to materialistic goals. To behave contrary to the truth means to cut the supply lines to the development of goodwill and from there to an unfair and evil human existence. As nothing is magic in our world, then the physical and spiritual machinery can benefit us and work properly if applied properly. Without practising, believing, and attaching ourselves to proper duties, ideas and religion will do more harm than good. Obviously, if we could cancel the clouds from the sky it could always be sunny and bright, but could life exist without rain? Can we succeed without proper religion? It could be less of a problem to not believe in anything, but can we survive with an empty mind? Most of the problems of today's world consist of the conviction that we can do without ideologies or proper physical exercise or any form of religion. Another mistake is the belief that there is no punishment for disobeying proper natural laws—call these laws ideologies; call them religion; call them love and considerations and physical motions. Our present misunderstanding is cutting the veins of those good natural supply lines, and as we can see, we have a human world in turmoil, with pains, disease, evil, murder, suicide, the emotionally unbalanced, ignorance, pollution, destruction, barbarians, selfishness, etc.

22

Our human emotions have only two ways to move, bad or good. The bad way, like weeds, comes automatically. The good way is a "selection" separate from the bad way, according to an intelligent and selective process. The good way is what we call justice and human civilisation. Unfortunately, we really do not know the process we have been given by nature in order to stabilize proper ways, emotions, and good feelings, and despite all the attempts and various ideologies we are getting nowhere.

First of all, we must abandon the belief that we can form an ideology without consulting and obeying proper natural laws. In other words, then, it is we who must follow the world and not the other way round. All of the ideologies that have succeeded from the beginning of civilisation until today have been based on nothing else but blind and unfounded wishes of people. As we get older and die and there is nothing we can do to become young again, we can come to the conclusion that who is really in charge of this world is not people, but nature. It is then illogical to pursue people's ideas without any scientific knowledge of what really forms our human body—physically, emotionally, spiritually, and including the Soul. We must not necessarily follow natural laws at all costs, but the problem is that as Mother Nature is in control of everything, then it is obvious that ideas of our own will never go too far. We have had, since the beginning of civilisation, many ideologies, but after a good start and illusion then everything collapses and delusions always follow. The reason for this is mostly because we do not know the methods of stabilizing proper feelings, work, emotions, sharing, caring, love, and consideration. Once sharing, caring, and love are absent from the human spirit, then the astray old spirit takes over by making it more comfortable to be powerful and be above others and then enjoy this feeling, also deriving pleasure from the misfortunes and pains and suffering of others.

Therefore, any form of ideology existing in any nation will always end with revolutions of some kind. We have had thousands of years of human history, and what really has happened was nothing else but revolutions and wars. After a war, people come to some sense, dialogues, considerations, and a new understanding, but as the years pass by the astray old human spirit will emerge, colliding with people in good nature, and from there comes revenge, revolutions.

Naturally, things will continue to be that way forever unless we apply proper methods. To apply proper methods, as I said, we have to abandon any ideologies formed by people's ideas without being aligned with proper natural conditions.

Analysing the world today, which is mostly founded on people's ideas, we can easily find incalculable problems. At the present time perhaps there is not a single human being on our planet who never got sick once in his life or didn't need a surgical process or need medical, psychiatric, or pharmaceutical help at least once. Our failure to understand our terrain's natural process has caused incalculable damage to the human race and the cases of death and sickness can no longer be calculated in thousands, but in millions worldwide. In fact, millions of people are murdered each year; there are millions of homosexuals. Millions are dying of cancer. Directly or indirectly, millions of children die of starvation each year. Rapes are in an incalculable number; so are assaults, blasphemy, corruptions, superiority complexes, and other problems of all kinds. Suicide rates are frightening. Many people in this world die before reaching old age or natural causes. As far as nature is concerned, because we could not begin with a well-developed intelligence, then our human race had to live with all the pains that are familiar to us today and across centuries and centuries. But our old pains, wars, etc., formed history. When intelligence will be fully formed and nature understood, it follows then that pain can be avoided. Our human race's civilisation had its beginning way back in the forests, and their existence was certainly not without pains, ailments, wars. But we have succeeded in becoming what we are. Reason then to believe that blind ways are a grossly unstable natural way to succeed. However, without an intelligent selection of proper ways our human existence can still succeed but will be painful all the way. Man-made ideologies can still succeed because strong and bad people can have advantages above those who by personality, through descendants and ideological factors, have developed some ailments and weaknesses that prevent their having a normal career, intuitions, ways to have substantial thoughts, work properly, etc. In the world of plants and animals they have not much possibility of success because of the limited uses of intelligence. Therefore, there is no consideration when the weaker one dies or is defeated and then the better and stronger animals will survive, so this natural selection is the only way to make species stronger and more alert. As things are, then, it is not necessary for us to select the good ways, because the bad ways are also successful but also painful. We have reason to believe that if no God came from the sky to give us comfort and peace when our human race was in problems and wars, we can draw the conclusion that

Mother Nature is good enough in her way. But nature has raw laws, and as we have intelligence to make nature more realistic, then it is bad if we let the bad ways enter our human spirit. It is no good to God to put our world in peace when we have a natural system that tells us how to do it.

Our present ideologies in the world are not based on or founded on the meaning of the words *sharing, consideration, understanding, proper religion*, etc. Therefore, there are vast spaces in our civilisation where "evil" finds a very comfortable home. To make things worse, our human behaviour can be transmitted to our offspring. In fact, many can easily notice that there is something wrong with the younger people. We should begin to put natural limits and boundaries on our minds and behaviours; otherwise there is the risk of developing a human spirit (emotion, wishes, etc.) too personal and at a dead end and with this not only creating suffering and stress when wishes will not come true, but also going on a collision course with good people. Deviant inherited behaviours can appear in offspring having a physical molecular apparatus away from true systems leading to many strange behaviours. Adults can become so selfish that as long as things go completely their way they can be happy, but if small things go contrary to their wishes, they will curse and argue. If the adults have ideological boundaries and limits not according to that good natural process, this can create nervousness and difficulty in growing within reasonable performances. This leads to one of the reasons why humanity is confused and astray. Following natural good offered conditions creates a spirit that is calculated and measured and with it we can live the entire arc of life in a well-balanced mood, with love, sharing, happiness, harmony, etc.

Therefore, then, if we want a fair civilisation we must use all of the ingredients that solidify our human foundations. As I said, in our world we have substances and also space. Not everything belongs to the world of solids or substances, but some things belong to the nonsolid as well. For example, we cannot form a good ideology by evicting the word *sharing*. In any ideology where the words *sharing* and *consideration* are nonexistent our human existence will have a missing link and from here will come imbalances, ailments, and pains. The fact that all the ideologies that we have today in the world are having problems is because our ideological foundations are not aligned to sustain the weight and geometry of proper assigned nature.

Imagine building a skyscraper above a foundation that is only a few inches thick. It eventually will collapse. What now is collapsing in our human race is the physical part of the human body as well as the spirit, which is tempered to be too weak, creating a mentality for wars, ailments, rapes, murders, suicides, ways to live easy, depression. Obviously, if love, light, air, vision, intelligence, dignity, sharing, and caring belong to this world that we have, logically then, when we are missing any of the above, our life can only be partly functional. In our world, as we lack magical factors, we can survive time only across generations of offspring, but this has the danger to pass on in them our physical and spiritual problems. If our offspring behave wrongly, then the family, the people, and the nations will suffer the consequences. An inherited wrong emotion and will of the mind is not easily dissolved, and an individual practising the wrong behaviour is forced to live an entire life with a handicapped will and emotions that seldom will find proper accommodation and satisfaction even among a rich society.

It is opportune today, then, to have many ideologies that, if people want, can be separated from the rest so we can live according to the geometry of our existing minds.

Our human world, yes, it could end up one day with only one ideology, but because we live in a world that always has to move and change, then we have to expect splits and ramifications from that said central and single ideology. Today it is too early in nature to expect one global ideology for mankind, because many people have a very limited knowledge of this world, causing personal behaviour to be imperfect and also causing the ideological and physical parts of offspring to veer off-course, creating in them a will and a mental resistance to fitting only particular ideologies. Logically, there are many people who have a feeling for the good things and application of morality, religion, etc., and obviously these people cannot live with any offered ideology but the one that fits and comforts their more appropriate existence. Attempting to keep all people under one ideology today and expecting love and peace is an impossible dream, because our human spirit is mostly formed with attrition and bad levels from one person to another and from here mistrust, shattered dreams, sad surprises, death, and destruction both physical and mental, result. Only one ideology can have a temporary appearance when human emotions, reasons, feelings, etc., are properly developed. Many people think that perhaps we can live without ideologies, but we must always keep present in our minds that we cannot live without a cause and being busy in the right direction fits well in that very well stabilized good course of Mother Nature. Rivers, lakes, etc., have to hold water. Otherwise the lands would be inundated, causing agricultural disasters, making it impossible to grow vegetables, grasses, etc. The many ideologies have to drain people's imaginations and wishes in order to create a motion of ideas making our human civilisation possible. As most people are born more or less with an already preestablished course either in the wrong or right, then the wrong will find their actions not well accepted by those who have a very well-balanced biological and mental behaviour and this creates many contrasts and hate.

It is then opportune that good individuals chose an ideology to fit their behaviours and wishes that they enjoy, separated from bad people and those who have gone astray. With a civilisation in this direction, it will not offer chances for people with an unnatural mentality to hide or use other people as fulfilment of their wishes, profit, feelings, etc. In this way, if they cannot use others for their gains or profits, they will be forced to embrace the good ways and thus civilisation.

23

Pain in our world always has forced people to find better and more comfortable ways. But when people can avoid pain by using other people for gains and comforts, then pain will not be sensed in most people, making them immune to please the unlucky. So the best civilised way is to have many ideologies for each sympathiser, so this will close the doors to the many who are matured contrary to reality with the danger of pursuing an ideology to fit only their conditions, which if it is unnatural will benefit no one. Looking to our planet map, we can easily see that there are many nations, races, continents, etc. Thus if a nation falls morally, there are others to stand. If a continent has a calamity of any sort, then another continent will remain. In case of wars, things cannot go into madness, because many will remain sane. If forests will burn, others remain.

Having one ideology can harm the physical balance, awareness, and mental development of humanity. People, since the dawn of civilisation, discovered superiority as a good feeling being above other fellow humans, as well as feeling important when others provide for their comfort. Eventually, having only one global ideology would be in contrast with all those habits developed when human civilisation was at its lowest point. Thinking without basing the various conclusions on natural reality, people will find our ideas only provide for a primitive civilisation. Eventually in the years to come there will be an ideological battle between personal ideas and the created realities of this world. It is my hope that reality will win.

The truth is that if we understand and follow natural conditions in the good ways we always will go a long and easy way, but if we do the opposite the road will be paved with pain and with lots of chances of a premature burial. With personal mentality away from that good nature we depart from that pattern where our body's molecules have found their ways in adaptability across millions of years. An unrealistic pattern of the ideological system leads to the killing and destruction of that natural engineered process that makes our human life feel happy and with that the sense of a strong, serene existence.

It is well understood that six thousand years of human civilisation is not enough to know everything of this world, and that is and remains a problem. Our human civilisation, which came out from the dark of the forests, was only aware of the material things but unaware of the possibilities of body molecules and the Spiritual and feelings' process. So comforts, riches, vanity, conquests, etc., dominated over reason and good sense, deviating from a proper spirit and religion. Our ancestors absolutely had no clues to really believe in an afterlife, nor could they see what possibilities our body's molecules have. This way the option for all was to get as much comfort in life, unaware of the consequences. These kinds of thoughts or human behaviour have always created more problems than cures and in the end provide a "sour taste" for all. Eventually, opportunism and selfishness, personal ideas, etc., without any comparisons to natural and Spiritual laws belong to that branch that only produces pain and as long as we follow these unrealistic ways we will remain in pain. Our human body is a mechanism created by a natural process, and it is hazardous to change any condition in nature. For then we will dismantle our own motor and will feel sick with mistrust, having sabotaged our good emotions and senses.

In our world it is very easy to destroy, but we do not have the smallest chance of doing better than what Mother Nature assigned us. It is then a stupid illusion to pretend to have a good humanity without having solid foundations in good religion and a variety of ideologies and by not abiding by the whole natural "mechanical" good process. A process gone astray will create automatically all kinds of pains (wars, revolutions, rapes, murders, unrest, ailments, mental problems), and this is and remains always the other side of the coin as long as humanity will live along these lines. In a few years we will enter the third millennium and with it perhaps will come the beginning of combining the material world with the ideological, spiritual as well the Spirit (Soul). As we are mortal beings, if we base everything on materialistic thought the day will come when wars and huge destruction will follow, because our human spirit will become heavy with pains without seeing an exit and from here frustrations and misunderstandings will lead to madness. It is good to outflank our materialism with all those requirements necessary for a well-founded human spirit. This is a necessary step, for it requires a well-based ideology founded around the words *sharing and considerations according to proper natural conditions*.

Another step to fortify our civilisation and our human spirit is taking up the belief in life after death. As I said, the belief in terrestrial things only, no matter how big our achievements are, cannot make us happy because our natural mechanism needs other (Spiritual) requirements applicable to it. In the end, we all die, and this is a "gap" that can only be "filled up" by the belief in a resurrection in Spirit after we leave this world.

Our Soul, then, is real, and it will become in the future one of the most intricate and difficult to understand realities, but also very fascinating and mysterious, and will keep humanity interested for thousands of years. As I said, it is not smart to wait for other civilisations from the stars to come to our world and explain things, because we have a natural plan that is working very well in the molecules of living things. Besides, no one can change it. If someone from the stars will tell us what to do, then perhaps the Creator of this world will be grossly undermined. Perhaps advanced civilisations will not be interested in knowing about our economy, because they have all knowledge concerning this, but as all the goods and gold of the world will not make anyone immortal, then it is obvious that an advanced civilisation on another planet will look for Spiritual information. If one day out of necessity other ways to travel interstellar distances will be invented, we also will not look for materialism but for Spiritual answers and clues to understanding things of the other side of the world.

As things are created on our planet nothing can succeed unless predetermined conditions are followed. This principle applies to everything. Good and bad ways, both seem to play by rules for the success and development of any form of life. Departing from the point that, in a way everything is eternal and indestructible, then the existence of "good" and "bad" is not a factor applicable to terrestrial life only, but is an existing part of eternity. Bad and good ways, then, have been in existence forever, and both ways are immortal. Obviously, then, there must be chances in our existence to choose either way, not only at a material level as we know, but on Spiritual terms as well.

Religions across past centuries have always advised us to choose proper ways, but we had problems understanding and our human world has always suffered. Many of us around the world still today believe that our human existence is only arranged substance to substance and after we die everything will remain on earth and we have no Soul to enter another existence in another world. But if this is the truth and substances could do everything on their own, why then do said substances need this immense space that we see in the first place? Logically, then, that substance in this universe could not do everything on its own and therefore life in matter indeed can have dual possibilities. As in our world everything has many purposes, meaning, then, this immense space that we see gives many possibilities, such as that chance for our earth to travel around the sun as well as light, energies, and spacecraft moving into space for what concerns material things. In the Spiritual world Souls can travel immense distances also because the chance to move does really exist (in space). The real difficulty now lies in the understanding, because our planet has a huge variety of living things and not everything can go to another world after death. But we have to understand why God created us in "space," the reason being that space suits the purposes of life, movements, and departure in Souls. This can tell us clearly that the chances for each one of us to another world can indeed exist but are not by magical events or it does not matter what we do but is a selection according to certain values and obedience that an individual can select for the purposes. As our Mother Nature is a mechanism, she certainly cannot have been lying to us when across millennia she has inspired us towards religion, which tells us to be good. To be good is a first step to move in the right direction for an afterlife.

Departing from that point that if we have been created or assembled as humans by many many parts (cells, organs, etc.), tells us again that our world is not magic. As we have no magical possibilities, then it is not magically guaranteed to have a Spirit for an afterlife unless we do determined things to accomplish it. This does not necessarily mean that we must be or feel good always, because certain pains, fears, etc., can be beneficial to human causes.

For example, there is the fact that as our human civilisation will advance into new understandings there we can add good things to our knowledge and progress, but we can also add evil things for any form of destruction. Fear of punishment when we do bad things has been across centuries the only method that has kept this humanity on an acceptable course. Otherwise we could have deviated into any form of madness. So this fear has been beneficial. If we become more and more dependent on our own personal will and understandings without any superiority to tell us what to do, then we need *a not man-made* fear, which does exist and is real as a punishment after we die. Who created this world was very well aware of the fact that without specific laws and fear this human world could not have reached very far.

In the future as our human civilisation will reach new heights we can also become bolder and it will be easier to achieve huge and total destruction if no fear of any kind will be on the horizon. For example, some of the nations can become so powerful and their weaponry so advanced that they will have no fear of breaking any natural, good laws or *any* laws and considerations, etc. Evil things can also come if certain nations will become so strong and will not have competitors, but as we have no contest at all if we compare ourselves to God, then any nation will back up "smartly" and be aligned with proper ways assigned by nature and by God. If there was no life after death and no punishment then our human race in the future would become so bold and insensitive and heartless that it would be better not to have been created at all.

As our world has been created, then all living things

have a beginning and an end. Since there is a beginning, all forms of life cannot be perfect at first. Life on our planet, for example, is more perfect today than it was millions of years ago. Mother Nature has her time, and five thousand years of human history is not enough for our human race to have a complete civilisation. There are many, many finishing touches needed for the building (civilisation) to be a complete structure to function on its own. When everything is fully understood, then we can all very well "adapt" in the offered natural world that we have. With changing ideologies not everything is without pains and problems, even if the changes are for the better. In other words, people who from generation to generation, because of misunderstandings and because we are beginners in that natural growth process, have learned a mixture of good things and also very bad, mystical, and ideological habits as well. In this century many ideologies have risen, but because the understanding of the world we have was not yet there, then either the ideology was followed by revolution or wars or in other cases people had that frustrating feeling that "something is missing." Misunderstandings, laziness, superiority complexes, blindness to natural and "Spiritual" laws, etc., high ambitions, noncaring and nonsharing, hate, abuses, ignorance towards natural understandings, etc., are ideologies people have always used. Logically, if the ideologies that we have today are based only on people's ideas without any comparison and understandings towards our natural created world as a whole, this creates in us an "evil" or a lump attached to the good nature as body, mind, and spirit that have to live or exist as "renegades" to the sane principle. In other words, either ideological or spiritual problems and most physical ailments (including many forms of cancer) are nothing else but human mentality astray from that engineered motor offered by our Mother Nature, or the world we have. A human body and mind away from that standard naturally offered engineering is also the principal cause of spoiling in offspring with a genetic "setup" that will cause them to be born pursuing those accommodations where Mother Nature has not much to offer. It is absurd to change ideologies with the hope of making everybody happy when our present behaviour and understanding exist and have existed on the wrong course. In other words, if we do want changes we cannot expect things to progress in us by magic or to be happy while remaining attached to the old ideological habits and misunderstandings and badly off-course. If changes are needed, we cannot carry old habits with the changes. Otherwise what really changes is the name of the president and the name of the ideology.

Logically, as human imperfections have spoiled much of that proper genetic natural engineering at many levels we cannot expect that people for the time being all enter the same ideology. Therefore, we have to apply to the good followers that proper ideology that makes them happy and proud. As people gone astray they can see the results of right and wrong as long as things do not go very far from that proper natural plan. It would be nice if we could understand and follow proper ways so we could all have one ideology and the good people could be set as an example so the people gone astray could return to proper ways. But at the present time certain habits, uncertain thoughts, and unrealistic feelings live and are part of our present civilisation and are difficult to change. According to this situation, we need ideologies to fit each individual according to what he can handle. Otherwise the good people, instead of being followed in their good example, will be nothing else but spoiled, ridiculed, hated, and exploited. A stray spirit is not able to follow proper natural instructions because it has no love and consideration and can only live by creating hate and destruction as the only ideological "food" that can put a motive on personal emotions. On the other hand, the "ideological food" for good people is sanity, justice, and morality and a good religion. This tells us that to have one only global ideology for all is nothing else but something that could be compared to an everyday civil war. The motivations that make life possible in a nonmagical world are friction or attrition, as long as they are useful to humanity. It could also have been a beautiful dream if we could live without work, but again, work produces motion, which in turn creates muscle cells, a necessity to move and to live.

Another beautiful dream could have also been to have the human race without wars and problems. Unfortunately, before intelligence the body was needed first, then, feelings, motivation, balances, awareness, etc. Then comes the actualization of intelligence in a primitive form, then the development gradually in full. Also, it would be beautiful if Heaven (as that "beautiful place" that we reach after we die) mentioned in most religions could be reached without doing much about it, but according to natural information (nature never lies, because it is a mechanism) things are not that way either. What we have to understand in reality is the fact that who created this world was not in control of eternity but had limited possibilities. Then we must abandon ideological grandeur, easy gains, magic, and big dreams out from that right, natural created plan/process and live according to what Mother Nature can offer.

From one process of understanding to another we can also conclude that all substances in space (stars, moon, planets) had to be in the constant presence of this "emptiness" (space) as a breaking point for infinitesimal small matter parts (atoms) as the only way to form variations of living creatures. Atoms in elements serve the purpose of forming different parts of our human body, for example. If all substances in space could have been indivisible, then they could only have formed something of their own or a "one unique piece" useless to create our natural world that we have and see, because we must have differences.

Divisions (atoms, molecules) then made possible our human body, which is a unity of little pieces of mechanisms made of substances, also is for the rest of living things. To keep functioning and alive a variety of living creates style, pattern, uses, and forms are needed. We can say the same about the need for many ideologies, as long as all the ideologies work to form a workable ideological mechanism. Otherwise we can reach certain conditions or stray behaviour where not much can be done.

In our human body we have many, many parts (organs), and all of them work to our advantage; so it should be for ideologies and for all the rest. Our earthly life is not Heaven, nor is it all roses, and not everything that nature offers is without conditions and a price. For one example, there is the heat of summer and the cold of winter, we have to work and raise a family, we get sick, we will become old, etc. All of this is not pleasant, but things can become more catastrophic, boring, painful, and crazy going contrary to the natural given condition. Religion and ideologies, like the sky above and the water and the air, will remain unchangeable, leaving the opportunity for people to choose independently the opportune portion suitable for their dreams. Evidently as intelligence accumulated gradually (because we needed a body first), the illogic, the unknown, misunderstanding, hate, and wars appeared among us frequently. This has also caused a growth out from that natural pattern of human feelings and emotions, creating terrible physical handicaps as diseases as well as spiritual problems. These are the causes of us having lost our proper behaviours, which now prevents us from behaving in fair conditions and harmony. At the present time, then, because centuries of misunderstandings have created hate and from here wars and a weak body and mind, then all these unnatural conditions of the human body and mind can overrule, damage, spoil, and dangerously tilt our civilisation as well those few where nature succeeded to let them be born "just" or straighten them properly during life. If one ideology will become the leader of this world, then logically, people under unnatural conditions will abuse any good person in order to have their ways and their illogical attachments and unfair wills.

As we have to develop intelligence, muscle cells, awareness, feelings, emotions, etc., then once again it can be proven that our life is a mechanism and needs certain conditions to develop. The small planet that we have, for example, because of its small size and limited conditions, is very useful for the actualization of intelligence across attrition and contrasts, wars, pains, and limitations. This principle applies also to all other living things. It is understandable, then, that if our earth's size was infinite then our intelligence, awareness, etc., could have taken a very long period of time to develop, because all those limitations and pains as the ingredients of developing the aforementioned awareness and feelings were not there in that normal amount required. (This does not necessarily mean that all planets in the stars are small, in order that limitations create intelligence, because if a planet is immense there could exist much nastier things than occur on smaller planets in order to speed up understanding, awareness, etc.) To create means to build. To build we need a blueprint and a starting point not entirely according to the will of the builder, because there are priorities to follow. (For example, we cannot build the roof of a building without the foundation.) Also, in our human life Mother Nature could not build intelligence in us before the body. This fact obviously created problems in the emotions and proper intelligence of people, for which we are still paying the price today. Then it is logical that our human body and mind are away from that proper natural plan, creating most physical and ideological ailments that we have today in the world.

As things are, then, we in this world very often can encounter people who have very limited muscle and mental will to work because their genetic body blueprints inherited from past generations lack in perseverance, natural proper conditions, misunderstandings, superiority complexes, etc. This mind and body conditioning led people to use others to perform work and thus these people feel superiority. For it is very difficult for stray human mentality to abide by any morality, religion, proper duty, or ideology, which explains, according to history, why no ideologies or religions or any other means have been able to keep this world's people in morality and peace. Perhaps in the distant future people will be more able to abide by ideologies and religion and from here progress to a more complete human success. Things like that cannot even be dreamed of today, because our human civilisation has come across or matured through distortions of all kinds, which have deviated from that natural genetic and mental spirit, building people with twisted feelings and unfair attractions towards proper ways. I repeat; it is opportune to let each individual of the world choose an ideology that he likes, and in this he can see by comparison and results or can learn through his own failures and pains a better way to succeed and survive.

The human race has emerged from the forest genetically and mentally unfit for any proper understanding, order, and ways and feelings. When the genes have taken this course in our human body, structures will be very hard to change in a short time, and due to this fact, at the present time we cannot put an entire humanity under one ideological system. For the good people whom nature has succeeded in making just will be overrun, cheated, and distorted by people born unjust. We do not have to expect that things always will be going for the worst, because Mother Nature's molecular plan does do the healing process as well, but unfortunately, natural healings are not always all roses but can be painful as well. Many times a wrong feeling can be buried by having a human ideology full of money, comfort, etc., but unfortunately, a bad spirit always comes afloat in minor ways during a good economy, with money and comforts, and badly in times of recession, poverty, wars, revolutions, etc. Our life on this planet is difficult from any point of view. Because of this factor, either evil, mistrust, or hardships develop, opening the door in most people to hating God, the church, and ideologies. For most people, then religion is laughable, a waste of time, and a weakness, while for other people, who have good considerations, love and understanding, etc., the church makes them happy, with serenity and attachment. So the dividing line is that a proper religion is what all humanity must have, but this cannot apply to those people where by misunderstandings and ignorance hate is stabilized in their feelings with selfishness, exploitation, wars, and immorality. Here the good ways and a good religion and ideology and Spirit are the *real enemies*. These astray conditions have to be lowered to proper ways in order to avoid a collision course with the natural, well-stabilized people and from here come problems.

24

Continuing in the search for clues, we can say that a structure or a creation has no eternal existence in its shape, design and conclusions. A mortal world such as ours is impossible to put at a stopping point at a level and time; then everything is based on a moving course. As a circle is the master or the principle of our geometry in all living things, then it is obvious that everything must circulate or "cycle." It is then nothing but an illusion and an unreality to persist in having one uniform ideology forever. As Mother Nature uses the same laws but with differences, we can then compare that to our planet, with its seas, lakes, and rivers to circulate and drain waters. It is then logical and natural to keep a few forms of ideologies that people can subscribe to or depart from according to their will, feelings, possibilities, etc., which in return creates joy and pain to keep life motivated, in circulation, fulfilled, balanced, reasonable, interesting, and going somewhere.

As already stated, immortals do not have to obey our laws, but mortals must under any condition obey proper ways and not the other way round. It is also absurd to accept bad habits as a natural condition, conquests of nations just to show superiority, etc., expectations of being famous forever or for too long, persisting in ideological and physical wars of any kind, grandeur at all costs, superiority complexes, etc. Persisting at all costs into certain habits and ways is not according to proper natural conditions but is a "deviant factor" or a "lump" growing against that proper engineering of the human body and mind. Again, as nature uses the same laws but with differences, we can note that in the growing process of plants is needed, for example, a balance between a mixture of hot and cold periods as one of those mechanisms for that growing and recycling process. If daylight, warmth, etc., were always exactly constant at all times, that would not have given to plants the right amount of reasons to form that process that generates that cycle of life, growth, and renewal. If plants, for another example, did not have that category of herbivorous animals that bugged them from time to time, then those plants had no reasons to build a defence (spines, poison, etc.). If the category of herbivorous animals had no enemy they had no reason to develop horns, muscle cells, awareness, and a form of intelligence.

Let's suppose now that certain nations want to stay ahead at all costs for a much longer period than the one stabilized by nature, then it will spoil in many the said natural proper linkages or those reasons that form our human growing physical and ideological process, such as: awareness, proper growth of muscle cells, ventilation of human motion across changes, proper stabilization of human behaviour, etc. If animals have formed defences (teeth, horns, claws, quills), it usually means that these were either for defence against enemies or some other necessity. The same laws apply. If a nation persisted with unfairness towards another, then armaments and perhaps wars will follow, telling us clearly that any motion by a thing creates reactions by another.

Human behaviour is formed, then, across motion, attrition, and will continue to be as long as this attrition will not pass the limits. Then we burn (stray behaviour). An "enemy" is used by nature as a form of friction and attrition in order to produce or obtain other laws or appearances that, speaking on the subject of animal world, can mean paws, quills, muscle cells, awareness, intelligence. As nature uses differences to advance, shape, and keep our ways, we then have to learn how to use, follow, and obey natural laws within limits offered. Eventually, as I said, habits genetically inherited are hard to get rid of for most people and a just and fair ideology that does not sustain them can become hated and very boring, so it is essential for the time being to have many ideologies to fit the various actual fair and unfair conditions of people. An ideology to go away or astray from that standard proper natural way will always produce pains, ailments, and problems, and anyone who follows each ideology has to be aware of the consequences. We certainly cannot accuse our Creator or other people when we mismanage proper natural conditions but must look to ourselves if we are the causes for our personal suffering as well the suffering of others.

High ambitions, selfishness, hate, carelessness, non-sharing accordingly, no respect towards a good religion, failure to believe in an afterlife, improper awareness, taking the easy way, improper teaching, lies, superiority and inferiority complexes, provocation without scientific reasons, failures to strengthen and follow proper requirements for our body's physical and ideological and Spiritual can really spoil the individual as well as entering and spoiling future offspring's proper genetic environment. Lack of understanding seems to be the "deficiency syndrome" that all present and past ideologies always have had. Everything can be considered a normal process for "beginners," as is civilisation on our planet at this point in history, but things certainly cannot remain idle forever at this point in history. As life is a circular geometry that is actualized by our earth around the sun and the moon and its phases, then everything is a cycle ending. As this is the only geometry that we have on earth, then all forms of life use only that same principle. This principle we call a cycle and it plays by the rules of life, new life, and death. A geometric circle involved in our life does not have the possibility of completely transforming or traveling forever, so chances are it will circle in the same place obtaining always the beginning and the end. A circle can be divided in sectors for other forms of geometry, which builds up the design of various vegetables and flowers and the rest of creatures' designs, then everything has a beginning and an end or is "cycled" and recycled accordingly. If we on this planet instead of having a circle for a geometrical figure for our living process had a "straight line" there could not be a chance of dying because of the impossibility of completing the said "cycle" or bent line. Unfortunately, matter or substances are not valued enough to form a world in a straight geometrical line and only Spirits or Souls can be the reality for this other or another world. According to most religions, we have chances to be there by following determined conditions. After all, if everything is eternal in one way or another, it is then very easy to understand that there is a door for a way out to another world from any situation on earth. The great magic and mystery of eternity is the fact that whatever we imagine and think is there somewhere, somehow.

Mysterious is life as we have it, because we came out from strange circumstances if we think of when earth had a temperature of thousands of degrees and then the development of life came from bacterial forms. These facts can enlighten us in the belief that everything is possible, but great knowledge is needed by "Someone" to make this magic mysteriously appear visible or intuitive or for us to be what we are. The Creator, or God, has used "matter" to incarnate one of the infinite wonders and mysteries of eternity as is life on the planet earth.

A Creator, or a God, is also a wonder and an explorer of infinity. Know-how and intelligence seem to be universal if we can note that it seems to be a mutual understanding between material intelligent life and other existing intelligent forms behind our universe, because nothing came to destroy or change those normal, natural plans that could throw everything into complete chaos. Besides, I also believe that their presence alone, which is completely different from us, could badly damage the whole of humanity in terms of emotions, ideologies, Soul, and spirit. It is understandable that the natural plan has to work as it is without interferences.

To substantiate this claim, God created a world where it is very difficult for other creatures on other planets or the stars and universal Spirits to be among us because of the enormous difficulty and differences involved and their dependence on living and existing in their habitat as either terrestrial, in the stars or ex-universal. Eventually after we die and then experience changes, then things can be accommodated to withstand the "place" or environments of other worlds in this universe and behind (which according to religion, as Mother Nature tells, is what we call paradise).

Returning again down to earth, as we have a circle as a geometric figure for all living things, then unless there will be a global understanding, then willing or unwillingly history will repeat itself across the things we already know: wars, starvation, diseases. All this will be progressed and presented from people with evil and astray behaviours. These said bad people will be the main causes of alternating and rotating in cycles all past ideologies from monarchies to dictatorships, from dictatorships to socialism, from socialism to freedom, from freedom back to a form of monarchy, then dictatorships, socialism, freedom, and so on.

If the foundations that govern our lives and the lives of all other living things on our planet are based on a circle, which is the only geometry available, it becomes obvious that we can only continue to move and live in the way our predecessors did in terms of life and death, war and peace, joys and pains, revolutions and hardships. So the same trash can will remain. However, by understanding, perseverance, obedience, devotions, love, etc., we can use any physical and spiritual law available on our planet for a better humanity towards better ideologies, religion, spirit, and Soul even if the foundations in the "trash can" are always the same. Nature from its plan always attempted to heal a stray offspring, the young ones and adults, but on the other hand, even a sane and good person can go astray when involved in an unclear act, boldness, provocations, superiority complexes, etc. Those deviating from the proper personal environment will affect future offspring. This seems to be that same unending circle that we have seen since the beginning of human history. Eventually we cannot keep only one ideology in this world. For then there will be always a collision course between followers and nonfollowers, leading to killings and destruction always on both sides. In order to stop hardships, wars, revolutions, and murders we must have a variation of ideologies to fit the conditions of humanity. We have to be on our own and cannot always expect miracles and magic.

* * * *

We are existing on a natural created course, and divine interventions can be rare. As everything has existed forever, our Creator does not reserve all the merits of

Creation and therefore simply does not have to come among us to show us a personal grandeur but can point His finger towards infinity, for our human race. Both God and man can wander in it for all time in eternity.